THE TITAN II HANDBOOK

A Civilian's Guide to the Most Powerful ICBM America Ever Built

CHUCK PENSON

Published by Chuck Penson, Tucson Arizona.
Printed in USA.

For information on purchasing copies of this book contact Chuck Penson at: wa7zze@gmail.com

All photos by Chuck Penson except as noted.
Front cover photo: March 13, 1964. Titan II missile N-30 lifts off for a test flight from launch complex 395C at Vandenberg AFB in California. (USAF photo)
Back cover photo: Missile N-10 waits in it's silo at the Titan Missile Museum. (Author's photo)
Author photo by Kathryn Bevacqua.

Book designed by David Steinlicht, Saint Paul, Minnesota. Book assembled on an Apple Macintosh using Microsoft's Word and Adobe's Photoshop, Illustrator and InDesign. Book text is Bitstream's News 706. Headlines and captions are Hoefler & Frere-Jones' Knockout.

First printing: September 2008
Second printing: November 2012

ISBN: 978-0-615-21241-8

For the men and women
of the United States Air Force
who served with Titan II.

Well done.

TABLE OF CONTENTS

INTRODUCTION AND ACKNOWLEDGEMENTS

Most of us, myself included, have never seen a real nuclear weapon, nor are we likely ever to see one. Nuclear weapons are, understandably, maintained with a certain degree of secrecy. So too, are the systems with which they are delivered. This makes knowing a lot about nuclear weapons something of a challenge for anyone outside of the relative handful of people we trust to look after such things.

Generally, one has to wait for years, sometimes decades, after a particular weapon system has been retired before civilians can know much about it. Even then, a veil of secrecy often lingers. The realm of nuclear weapons is both arcane and secluded. Admission is nearly impossible. Yet now and then something truly remarkable happens—one of them gets turned into a museum.

Such is the case with Titan II. And Titan II's story is even more remarkable than most. On the day the Titan Missile Museum took its very first group of tourists into a launch control center near Tucson, Arizona, other Titan II control centers in Kansas and Arkansas were filled with majors, captains and lieutenants still patiently waiting for the order to launch their missiles. As astounding as it may seem, while sergeants and airmen in Kansas and Arkansas were still reciting six-letter codes to gain access to their launch sites, in Tucson tourists laughed and joked and snapped pictures of each other turning the launch keys. Who could have imagined such a rapid and astonishing turn of events?

Many of the people who contributed to this book have seen nuclear weapons—up close and personal. They have spent many years of their lives deep inside the closed and cloistered world of nuclear weapons. Many have spent years with their fingers on the trigger.

In one of those rare and serendipitous turns of fate, a few of these extraordinary people retired from service at the very moment the Titan Missile Museum was being born, and glided smoothly from a top secret world into one thrown wide open to the light of day. Lt. Col. Orv Doughty USAF (ret.) and Col. Hugh Matheson USAF (ret.) are two such individuals. These two officers, whose service dates from the very beginning of the nuclear age, instantly grasped the significance and importance of a museum devoted to Titan II and together they took up the cause. They gave briefings, talks and presentations promoting the museum and its mission. They met with their still-uniformed colleagues, finagled precious artifacts and cut through miles of red tape. Once the museum's doors opened they wrote training materials, taught docents about Titan II (all the while being careful not to divulge anything secret) and provided most of the human thrust required to get the museum off the ground. It is largely from their dedication and determination that we now enjoy such a rich and detailed picture of one of the most important weapons systems ever deployed by the United States. We owe them a debt of gratitude.

◄ Standing guard at the entrance to the headquarters facilities of the Strategic Air Command, Offutt Air Force Base, near Omaha, Nebraska. Circa 1959.

NATIONAL ARCHIVES VIA NATIONAL SECURITY ARCHIVE, GEORGE WASHINGTON UNIVERSITY

Twenty years and a million tourists later their initial enthusiasm is still largely intact, and they still love to talk about Titan II. Both have answered hundreds of my questions and provided clarification on dozens of critical details. They kindly agreed to read over the manuscript and made a number of very helpful corrections and suggestions. Lt. Col. Dwight Mears USAF (ret), Lt. Col. Chuck Smith USAF (ret.), Major George Birch USAF (ret.), Major Bob Williams USAF (ret.) and retired Air Force civilian communications expert Brad White also read the manuscript, answered many questions and offered important insights into the inner workings of the Air Force in general and the Titan II missile program in particular. Additionally, former Air Force captain and Missile Combat Crew Commander Yvonne Morris was instrumental in my understanding of Titan II.

A number of civilians also added immeasurably to the book. Arnold Applebaum granted unrestricted access to several former Titan II missile sites he owns. Many important details in the book would have been omitted without his gracious cooperation. Neil McSweeney was exceedingly helpful in checking my math and the many unit conversions required to tell the story. My thanks also to Nick Halliday and Jane Smith. They offered to read the text and made numerous corrections and helpful suggestions. And Dave Becker did an absolutely superb job digitizing dozens of 35mm Air Force training slides.

I would like especially to thank Leslie "Pez" Owen for her tireless efforts in proofreading the book. She has read the text more times than anyone other than me, and her helpful hand is evident throughout the work. Additionally, Ms. Owen's piloting skills made possible an aerial survey of all 18 Titan II sites around Tucson, and hundreds of aerial photos, some of which appear in the book. Her assistance with my research on the ground at site ruins was also extremely helpful.

Teresa Vanden-Heuvel, who works with the Aerospace Maintenance and Regeneration Group (AMARG), Tucson's famous aircraft "boneyard," gave me unimpeded access to the facility and spent several hours under Tucson's merciless sun with me as I photographed, measured and studied the remains of Titan II missiles then in her care.

Dr. David Stumpf, the world's foremost civilian authority on Titan II, gave me more of his time and energy than I could have imagined possible. Numerous key points and details were saved from misrepresentation thanks to his efforts.

The National Security Archive maintained by George Washington University contributed several wonderful photographs rescued from the National Archives in Washington DC. George Washington University is to be commended for its unrelenting use of the Freedom of Information Act in pursuit of contributions to public policy and public understanding of many government programs.

Graphic designer and long-time friend David Steinlicht did the layout and design of the book. Nicely done, David.

It is fair to say that this book would not have been possible without the cooperation of the Titan Missile Museum, located about 30 miles south of Tucson, Arizona. The museum provided unrestricted access to its archives containing tens of thousands of documents and photographs related to the Titan missile program. My deepest thanks to Deputy Director Yvonne Morris, her staff, and all of the docents who every day so gladly sacrifice their time and give freely of their knowledge in the name of enhanced public understanding of nuclear weapons, and the vagaries of deterrence and defense. They are engaged in a noble endeavor.

On a personal note I would like to acknowledge the profound influence of my father on my life in general and on this book in particular. Sadly, he did not live to see the finished product, but his fingerprints are everywhere here. He kindled within me a keen sense of wonder and imagination, and instilled in me the idea that anything is possible. Thanks, Pop.

Lastly, I would like to thank Kathryn Bevacqua, loving wife, faithful friend, trusted companion and my only true joy. Every day with you is a wondrous new adventure. We are the power of two.

While all these individuals did their very best to make certain every "i" was dotted, every "t" crossed, every fact correct and every reference sound, I alone am responsible for any and all errors which may be discovered among these pages.

A number of people have asked me to explain my interest and enthusiasm for something so esoteric as a nuclear weapon system. They are curious as to why I would devote so much time to the investigation of Titan II. It is not specifically Titan II that interests me. It is, as physicist and Nobel laureate Richard Feynman once remarked, the pleasure of finding things out.

To everyone who has contributed to this project and assisted me in finding things out, I offer my sincere, humble and heartfelt thanks.

Chuck Penson
December 2012

NOTES ON THE TEXT

All the information in this book was compiled from unclassified or declassified documents, the author's field research, and from information in the public domain both in print and from on-line sources.

◀ SAC's global headquarters at Offutt Air Force Base. The three-story administration building is connected by underground ramps to a three-story control center 45 feet beneath the surface. Circa 1958.

NATIONAL ARCHIVES VIA NATIONAL SECURITY ARCHIVE, GEORGE WASHINGTON UNIVERSITY

The archives of the Titan Missile Museum were of particular importance.

Throughout the book nominal or approximate values have been used for numbers associated with various aspects of Titan II. In addition, most numbers have been rounded for ease of discussion. Values cited are based upon the most credible sources available, and from manufacturer or military specifications when possible.

In most cases it is neither useful nor productive to dwell on exactitude. More often than not the precise value of a number is not known, and cannot be known. For example, even the manufacturer did not know the exact weight of the silo closure and blast doors, and it has not been possible to put them on a scale. Indeed, very few numbers associated with Titan II can be known with any precision. The weight of the missile, flight time, staging times and apogee, for example, were different for every missile because they all depended on the target. The yield of the warhead, generally agreed upon to be 9 megatons, has in fact, never been declassified.

Docents at the Titan Missile Museum routinely say that the missile could be launched in just 58 seconds. This too, it turns out, is imprecise. Actual launch data reveals the time varied from about 40 to a bit more than 60 seconds. For simplicity, and because the number is so widely used, this guide will use the 58 second figure when referring to launch time.

Those in aviation, and by extension those in missiles, refer to distances almost exclusively in nautical miles. This arcane unit of measure is meaningless to virtually all civilians and all but a handful in the armed forces. In the name of clarity I have, therefore, risked the wrath of some by choosing to state distances in this book in statute miles (and kilometers) — the same unit of measure ticked off by the odometer in your car.

In the interest of simplicity and brevity I have in many cases generalized the discussion of otherwise highly detailed and complex systems and procedures. In doing so I have attempted to remain faithful to the most salient points. Most of these generalizations will be undetectable to anyone but former crew persons, and I apologize to them in advance.

This book will present operational and technical aspects of Titan II as they were defined in 1982, the year launch complex 571-7 went off alert. Differences both subtle and stark may be noted by crew members on alert after that date, and by those who served significantly earlier.

Except in the most general terms, this book does not discuss targets. Titan II's targets were classified and unknown to the crews, and remain classified as of this writing. It is generally accepted that targets included a wide variety of military and industrial facilities across the former Soviet Union. With respect to China, it is important to understand that Beijing (referred to as Peking for much of the Cold War) lay at, or perhaps just beyond, the extreme limit of Titan II's range, and that virtually all the rest of China was well out of reach. Further, there is no credible evidence to suggest that Titan II ever targeted Cuba. In fact, the TRW Space Technology Laboratory Final Report on Titan II, dated 1965, makes clear that limitations of the guidance system put Cuba out of play. And while a guidance system update in 1979 would have made it possible to strike Cuba, doing so might have been ill-advised. Detonating a 9-megaton hydrogen bomb over, say, Havana, would, at minimum, have flash-blinded tens of thousands of Americans in southern Florida and across the Deep South. The United States had a number of other nuclear options it could have deployed against Cuba, had it ever come to that.

All highly specialized endeavors have their own language and the military, with its myriad esoteric programs, has embraced jargon and acronyms with a passion. In order to talk intelligently about Titan II and in order to understand those who talk about it, it is essential to know and understand the specialized words and acronyms that accompanied the missile's operation and maintenance. I have therefore not shied away from using Titan II's unique terminology and have attempted to explain it clearly when it occurs. A glossary has been included for quick reference.

With apologies to all who know more about physics than me (and that would be just about everyone), I have greatly simplified the discussion of the workings of nuclear weapons. In leaving out many details I have tried carefully not to misstate the facts that seemed, in my opinion, most relevant. Those interested in a more thorough and competent treatment of nuclear physics are referred to the bibliography.

◀ Underground at SAC Headquarters. Large screens present information on weather conditions, force deployment, aircraft and missiles, plus a multitude of other operational data. Circa 1961.

NATIONAL ARCHIVES VIA NATIONAL SECURITY ARCHIVE, GEORGE WASHINGTON UNIVERSITY

In October, 2012, just as this second edition was going to print, most of the topside propellant handling equipment (trucks, tanks, etc.) at the Titan Missile Museum (and pictured in this book) was removed from the complex for restoration. Some pieces are slated to return late in 2013.

Statements and opinions expressed in the book do not necessarily reflect the views of the Arizona Aerospace Foundation or the Titan Missile Museum.

Finally, please permit me to address a pet peeve. If you want to be taken seriously in a discussion of nuclear weapons, you would be well served in pronouncing the word correctly. It is "nuclear" (NEW-clee-er), not "nucyaler" (NEW-kya-ler).

FOREWORD

Nuclear war is a messy business. Consider the possibilities: entire cities burned off the face of the Earth, hundreds of millions dead, many more injured, many permanently disabled, lingering radiation, rampant suffering, starvation and disease, long term increases of cancers and birth defects, perhaps even nuclear winter. No matter how you look at it, the world you knew before the first bomb exploded would be gone. Forever. To paraphrase the late British geneticist J.B.S. Haldane, not only would the devastation be worse than you imagine, it would be worse than you *can* imagine.

Is that what Titan II and its counterparts in the former Soviet Union would have wrought? Perhaps. But Titan II was not designed primarily to fight a nuclear war, it was designed to prevent one from starting.

Titan II was conceived as a retaliatory weapon, to be used only in response to an attack. Its fast reaction time and enormous destructive power made clear that if we were forced to retaliate, the consequences for the Soviets would be so unspeakably horrible that perhaps they would prefer not to pick a fight with us in the first place. This unnerving strategy was referred to as deterrence.

Dr. Strangelove said it best in Stanley Kubrick's 1964 classic film *Dr. Strangelove: Or How I Learned to Stop Worrying and Love the Bomb:* "Deterrence is the art of producing in the mind of the enemy, the fear to attack." Indeed, a little fear can go a long way.

During the Cold War most Americans (and much of the rest of the world) thought about nuclear weapons a great deal. The idea that the end of the world could come at any moment loomed large and created within many, a fear that was not only real, but also occasionally palpable. During the coldest moments of the Cold War, during the Cuban missile crisis for example, many people built, or seriously considered building, fallout shelters.

As we drift further and further from the Cold War years, Americans, at least, find themselves thinking less and less about nuclear weapons and the possibility of nuclear war. The fear we once felt has largely dissipated. This is not necessarily a good thing. For many born after, say, 1980 or so, nuclear weapons are a mere abstraction, thought of, if they are thought of at all, as perhaps nothing more than large bombs, and that one could hope to survive a nuclear attack by hiding in the basement or under a desk. This is a dangerous misunderstanding, equivalent to believing one could hope to survive a 75 mph (120 km/h) head-on collision by bracing oneself against the dashboard.

This book is intended to acquaint the reader with the apparatus of nuclear war and the complexities required for its deterrence, or failing that, its execution. It is intended to persuade the reader to muse for a few moments on the consequences of a nuclear exchange, that nuclear weapons exist, that thousands of them could be launched in a matter of minutes, and that nuclear weapons probably pose no less a threat today than at any time during the Cold War. This book is intended for researchers interested in Cold War history and technology, for armchair tourists, and for those contemplating a visit to the Titan Missile Museum or one of the many other artifacts of the Cold War.

In an act of great foresight, a group of dedicated and thoughtful civilians, working with the cooperation of military officials during the time of Titan II's deactivation, devised a plan to save from demolition a complete Titan II missile and launch complex: site 571-7, located just south of Tucson, Arizona.

The Titan Missile Museum and National Historic Landmark now preserves the only complete nuclear weapon system of its kind in the world, and is perhaps the ideal classroom in which to gain a fuller appreciation for the intricacies involved in defending the United States and the awesome responsibilities of doing so with nuclear weapons.

CHAPTER I

A NUTSHELL HISTORY OF THE ICBM AND TITAN II

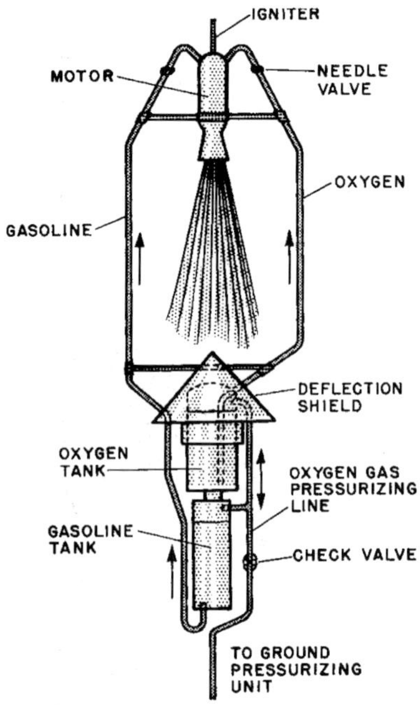

▲ Diagram detailing the design of Goddard's early rocket engines. Modern rocket engines work in much the same way.

Detailed histories of the intercontinental ballistic missile (ICBM) in general and Titan II in particular have been told in a number of books and websites and need not be repeated here. Still, a brief historical overview will be helpful in a fuller appreciation of both the machine and the complex that kept it ready to launch.

Although research into the science of rockets can be traced as far back as 13th century China, serious rocket science in the United States didn't begin until the early years of the 20th century.[1] Most notable among those working with these little-understood machines was Robert Goddard. Goddard was the first to develop and successfully fly a rocket using liquid fuel. All early rockets used solid fuel propellant much like the black powder used in fireworks. Liquid fuels were important because they contained far more energy per unit weight than solid fuel, enabling them to lift heavier payloads and propel them to greater distances. Goddard also was the first to successfully demonstrate gyro-stabilization, a technology used to keep a missile on course. Unfortunately, Goddard's secretive nature made it difficult for others to learn from, and expand on, his work, ultimately slowing the pace of rocket development in the United States. Goddard's *modus operandi* may also have impeded his own work by precluding the feedback so valuable in all manner of research. Although he invented and patented many important technologies, Goddard was never able to integrate all his advancements into a single, fully successful design. His best rocket, fired in 1937, reached an altitude of only 9000 feet (2.7 km).[2,3]

The real advances were happening in Germany. During the 1930s, groups of amateur rocket enthusiasts designed and launched dozens of small vehicles, learning a great deal in their casual way. To a person, they were interested in rockets as a means for space travel—not as weapons. While many, perhaps most, of the members of these groups were involved primarily for fun, there were serious investigators among them. Still, enthusiasm notwithstanding, these groups lacked the organization and money required to make any serious progress.[4]

When the German army began to investigate missiles as a weapon not banned under the treaty of Versailles, it looked to those enthusiasts for the talent needed to help get its research off the ground, so to speak. One who immediately caught the eye of army officials was Wernher von Braun. Von Braun was hired as a civilian employee of the army and ultimately developed

▶ Dr. Robert Goddard, American liquid-fuel rocket pioneer, circa 1940.

NASA ARCHIVES.

NASA ARCHIVES.

▶ Goddard and one of his early rockets. The propellant was gasoline and liquid oxygen. Circa 1915.

What's the Difference Between a Rocket and a Missile?

The difference can be subtle. Even the Titan II *missile* used *rocket* engines.

A missile is any object thrown, shot, or propelled toward a target. The bullet from a gun is a missile. Even baseballs are missiles.

A rocket, on the other hand, is generally thought of as a vehicle specifically propelled by the thrust generated from rapidly expanding gases, created by the burning of some form of fuel and expelled through a nozzle.

In general terms, rockets and missiles can be defined by their intended use. If the machine is used to deliver an astronaut or satellite into orbit, it is a rocket. But if the same machine is used to deliver a bomb to a target, it becomes a missile. Titan II was used for both purposes.

NASA ARCHIVES.

the V–2: a liquid fueled missile and the world's first fully successful long-range ballistic missile.[5,6,7] The V–2 was deployed too late in World War II to become a decisive weapon, but more than a thousand of them terrorized England and Europe during the final months of the war.[8]

Upon Germany's surrender, the United States captured von Braun and most of his rocket team. They were brought to the U.S., along with all the V–2s and technical documents that could be found, and were put to work developing rockets and missiles for the U.S. Army. It is fair to say that the V-2 paved the way for virtually all future rocket development in the United States and elsewhere. Virtually all rockets and missiles can trace their lineage directly or indirectly to the V-2.

Having seen the effectiveness of the V-2, the U.S. military recognized the value of ballistic missiles, but from 1945 to 1950 there was very little progress toward a truly functional ICBM. There were a number of reasons for this: infighting among the Army, Navy, and Air Force over which should take charge of this new technology; ever-changing policies of the Pentagon; extreme conservatism within the Truman administration with respect to ballistic missiles; the Air Force's unwavering commitment to manned bombers; and skepticism among high placed experts that ICBMs were technologically feasible. Indeed, scientific luminary Vannevar Bush (inventor of the analog computer and Director of the Office for Scientific Research and Development during WWII) stated that in his opinion an ICBM would not be possible "for a very long period of time to come."[9] Furthermore, secure in the belief that it had a firm monopoly on the atomic bomb, the United States did not see any need for haste in developing a ballistic missile. The best estimates were that the Soviets would not have an atomic bomb for at least ten years and that when it finally did have one it would be delivered with an airplane.[10,11]

NASA ARCHIVES.

▲ Wernher von Braun with a model of his V-2 missile, circa 1950.

▶ Von Braun's V-2 paved the way for American rocket development. Circa 1950.

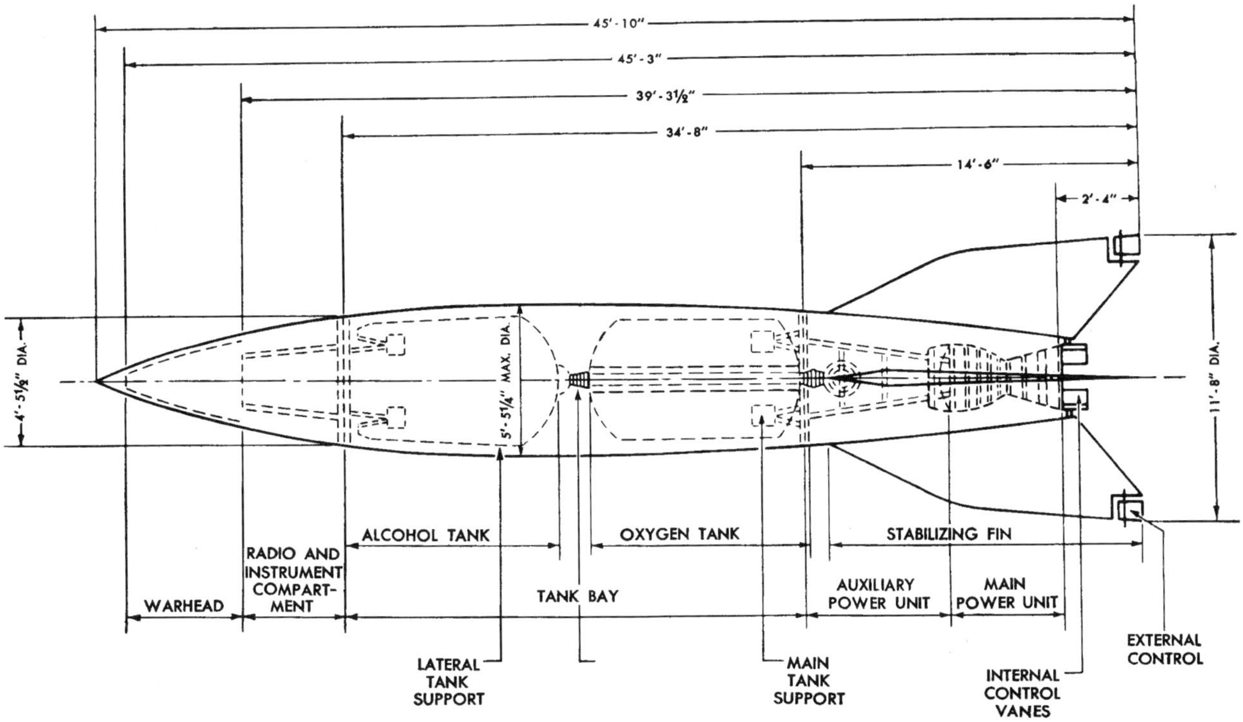

NASA ARCHIVES.

In the Soviet Union things were very different. Missile development remained firmly under the control of the Army well into the 1950s, and there seemed no doubt about the feasibility of ICBMs. With the help of captured "B Team" German rocket engineers, the Soviets had restarted the V-2 assembly lines, building and testing ever larger and more powerful versions of the missile under the direction of Sergei Korolev, Wernher von Braun's Soviet counterpart. From the beginning, Korolev's emphasis was on "heavy launchers"—rockets powerful enough to deliver an atomic bomb to the continental United States.[12]

◀ Soviet rocket designer Sergei Korolev. His powerful R-7 launched Sputnik—and the space race. Date unknown.

NASA ARCHIVES.

Three events finally pushed the United States out of its missile lethargy: the fall of China to communist insurgents in 1949, the invasion of South Korea in 1950 by communist North Korea (which, incidentally, was armed with the latest in Soviet weaponry) and, perhaps most disturbing of all, the Soviet Union's successful detonation of an atomic bomb in 1949, years ahead of expectations. These events spurred a variety of military projects, among them an underfunded program to develop an ICBM capable of delivering a nuclear warhead to the Soviet Union. The project was code named "Atlas."

One of the main problems with the concept of nuclear tipped missiles was the weight of the warhead compared to its yield. The first generation of atomic bombs weighed as much as 8000 pounds (3600 kg) yet had yields of only 20 kilotons or so. This meant the missiles used to deliver them would have to be huge *and* have pinpoint accuracy. Huge missiles would be very expensive, and pinpoint accuracy was simply not possible given the technology of the day. As originally conceived, Atlas would use five or seven engines, stand 160 feet (49 m) tall and weigh in at 440,000 pounds (200,000 kg).[13] Critics argued that missiles were not a cost-effective means of delivering nuclear bombs, and the program languished.[14]

But the troubling development of the Soviet atomic bomb also spurred the U.S.

◀ The Atlas missile, the United States' first operational ICBM, shown here in a test launch from Cape Canaveral. Circa 1958.

NASA ARCHIVES.

into developing what physicists called the "super"—the hydrogen bomb. The proof-of-concept for this terrifying weapon came in the form of a 65-ton (60-metric ton) experimental device using a form of liquid hydrogen. Code named "Mike," the device was detonated in 1952 on a small atoll in the South Pacific. It exploded with a force equal to ten million tons of TNT—nearly 700 times more powerful than the first atomic bomb, and the single most powerful explosion on Earth up to that time.[15] While Mike was not a deliverable weapon it proved that a

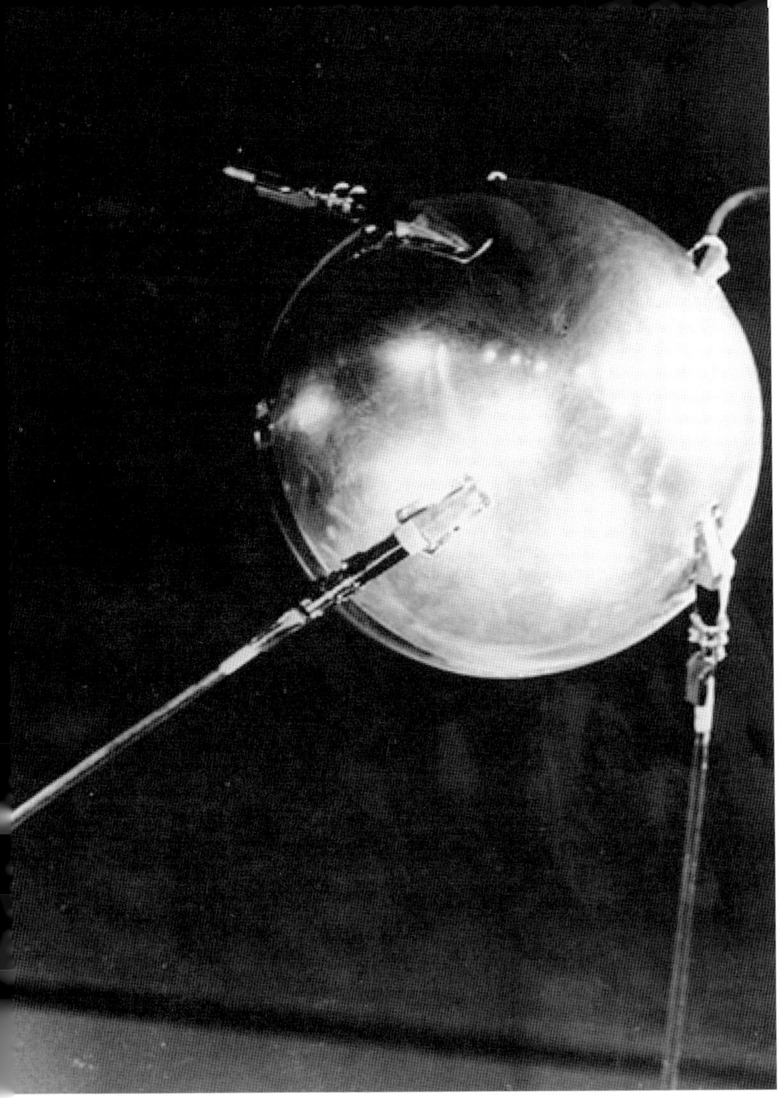

NASA ARCHIVES

◀ Sputnik, the world's first satellite, was easy to see crossing the night sky over the United States.

new generation of lightweight but extremely powerful bombs lay just around the corner. The promise of lightweight, high-power hydrogen bombs breathed new life into Atlas, but two other events pushed the program into high gear. First, in 1953, the Soviet Union detonated its own hydrogen bomb. Second, that same year U.S. intelligence agencies found evidence that the Soviets were far ahead of the U.S. in missile development and might soon be able to deploy nuclear-tipped missiles in numbers large enough to permit either nuclear blackmail or a pre-emptive strike.[16]

Indeed, the Soviets had been making a lot of progress. Korolev had made a series of incremental improvements to the V-2 engine, scaling up the machine, making it larger, and increasing its power. These changes would allow the missile to travel further and carry a heavier payload. By the mid 1950s Korolev had settled on the design of a huge missile designated the R-7. The missile employed almost three dozen small engines grouped into five clusters. All five clusters got the vehicle off the ground. Then, when the heavy lifting had been done, four fell away, leaving the main booster to finish the ascent. Korolev chose this strategy because small engines were intrinsically easier to make and more reliable in operation, but the design had its challenges. Feeding multiple thrust chambers and making sure they all ignited and continued to operate for the duration of the flight was not a simple task—even for a rocket scientist. The R-7 was not without its failures, but Korolev persisted. In August 1957 the first successful R-7 lifted off and after a perfect flight splashed down in the Pacific Ocean 5000 miles (8000 km) downrange—just far enough to demonstrate that it could hit the United States.[17]

In the U.S., engineers went back to the drawing board and redesigned a much less expensive Atlas around the premise of a lightweight warhead. The missile shrank from 160 to 75 feet (23 m) while the weight fell by half, to 220,000 pounds (100,000 kg). Moreover, it would now require only three engines. And perhaps best of all, high accuracy would not be required: hydrogen bombs were so powerful it wouldn't matter if they hit their targets precisely.

Over a period of several years, six versions of Atlas were built and tested. Versions A, B, and C were all developmental missiles flown exclusively to refine the basic systems. Flight tests began in June 1957. There were many failures, including a couple of spectacular explosions prior to which the missiles spun loops, but eventually successes began to mount.[18] Atlas burned liquid oxygen (LOX) and RP-1 (a highly refined kerosene), and was referred to as a "stage and a half" missile, similar in operation to Korolev's R-7. While later missiles would have a second stage to take over once the missile had gained significant altitude, all of Atlas' engines shared a single set of fuel tanks. All three engines ignited at lift-off, then during the flight, much like the R-7, two engines dropped away leaving only one to complete the ascent. The stage-and-a-half strategy was chosen because of the difficulty in igniting an engine at high altitude.[19]

Then, on October 4, 1957, the world was electrified by news that the Soviet Union had launched the first Earth-orbiting satellite—a technological achievement of stunning proportions. "Sputnik" weighed about 185 pounds (84 kg) and was lofted into orbit by Korolev's R-7. Sputnik's technical accomplishment was momentous, but its most significant impact was political and psychological. The Soviet Union could boast of having an operational ICBM, and had scored another first.[20] In Washington, news of Sputnik initiated much hand wringing, gave rise to the term "missile gap," and vaulted

▶ An Atlas F, venting oxygen, rises from its silo during a propellant loading exercise. Circa 1960.

USAF PHOTO. TMM ARCHIVES.

the United States into a "space race" with the Soviets. It is no exaggeration to say that Sputnik may have done more to hasten the development of the U.S. ballistic missile program than any other single event.

After years of indifferent funding, false starts and cancelled programs, Atlas D became an operational ICBM in January 1959—just fourteen months after Sputnik. The first operational squadron of six Atlas Ds was pressed into service at Cooke Air Force Base (now Vandenberg Air Force Base) in California. Three of the missiles stood on unprotected launch pads, a fourth lay in a horizontal launcher and two others waited in underground silos.[21] The Air Force may have deployed the missiles mostly for show—Atlas D was still largely in development and whether it would actually have worked if called upon to do so is still the subject of debate.[22] Ultimately, however, Atlas proved reliable enough to also be used in Project Mercury, NASA's first program to launch astronauts into Earth orbit.

Over the next three years, improved Atlas E and F missiles went "on alert" at 11 other bases. The last group was installed in 1962 at Plattsburgh AFB in New York and was the only group of ICBMs ever to be installed east of the Mississippi River (excluding Snark).[23] At the peak of their deployment, 123 Atlas missiles were on alert, but in the end, Atlas stood guard for only six years. The last groups were deactivated in 1965.[24]

In 1955, while Atlas was still under development, the Air Force authorized the development of a second missile system. It did so for several reasons: to hedge its bets in case of problems or delays with Atlas, to stimulate innovation and competition, and to ensure against complacency within a lone contractor. This new missile would be named Titan (Titan I). Titan could be developed with only a modest increase in cost since Atlas had already scaled much of the learning curve.[25,26]

Like Atlas, Titan burned LOX and RP-1. But Titan was a true two-stage missile, and at 90 feet (27 m) Titan was about 15 feet (4.5 m) taller than Atlas. Titan was equipped with an improved guidance system and was kept partially loaded with propellant. Although it was stored in an

USAF PHOTO, TMM ARCHIVES.

▲ A Titan I in launch position at Vandenberg AFB, circa 1962.

underground silo like Atlas F, Titan also could not be launched from within a silo. Just before launch the remaining propellant was loaded while still underground, then the missile would be raised to the surface and fired. Keeping Titan partially loaded reduced the response time of the missile, and keeping it underground until just before launching reduced Titan's exposure to enemy attack.

The first successful test flight of a Titan I missile was from Cape Canaveral on February 6, 1959.[27] The first Strategic Air Command (SAC) squadron was deployed at Lowry AFB near Denver, Colorado in April 1962. In total, 54 Titans were placed on alert at five locations in the United States. Improvements notwithstanding, Titan I's reign was even shorter than that of Atlas, lasting less than three years. The last squadrons were deactivated in February 1965.[28]

Both Titan I and Atlas shared a fundamental flaw that rendered them ultimately unsuitable for their intended purpose: they relied on liquid oxygen (LOX). For an ICBM to be truly effective as a deterrent or retaliatory weapon it must be capable of rapid response, and the use of LOX made that impossible. While there was no difficulty in storing the kerosene-like fuel component of the propellant, the liquid oxygen oxidizer was a completely different matter. At –290°F (-180°C) LOX boiled away very rapidly and therefore could not be stored aboard the missile: it had to be loaded just before launch. Loading LOX was a process that could take tens of minutes (on a good day), and in a nuclear war time is not on your side. ICBM travel time from the Soviet Union to the United States was (and still is) about 30 minutes.

Then there was the problem of launching: while Atlas F and Titan were deployed in underground silos, neither could be launched from underground.

These deficiencies were not lost on the Air Force and by the late 1950s officials began to explore the idea of a missile that could be launched from within a silo and in which both the fuel and oxidizer were storable for long periods.

The Air Force looked to the Navy for a solution. Both the Navy and the Army had been researching storable liquid propellant since 1945 and by 1958 they had found a combination of two chemicals with all the right characteristics. Specifically, they were fairly easy to handle, relatively stable, not excessively corrosive, and storable for long periods at normal temperatures. To the Air Force these qualities sounded like a winning combination. Accordingly, bids to build Titan II were solicited, and in July 1958 the Glenn L. Martin Company got the job. Martin built a brand new facility near Littleton, Colorado specifically for the task.

Titan II turned out to be the largest ICBM ever deployed by the U.S. It stood 103 feet tall and its stage one engine generated 430,000 pounds of thrust. That's more than 100,000 pounds greater than Atlas or Titan I, or if you prefer, just a bit less than two 747s running at full throttle.[29]

The first test of a Titan II took place at Cape Canaveral in March 1962. It was a flawless flight. The first in-silo launch was conducted in February 1963. The missile successfully cleared the silo, but was destroyed at 18,000 feet (5.5 km) due to a failure of umbilical cables to properly disconnect.[30]

It would take 21 more test launches through 1964 to work out all the details, but the Air Force didn't wait for the conclusion of test flights before it started construction of launch facilities. Work on the silos began in Tucson, Arizona on December 9, 1960, more than a year *before* the first test flight. Almost exactly three years later, on December 29, 1963, the last of 54 Titan II launch sites in three states (Arizona, Kansas, and Arkansas) went on alert.[31]

The Titan II program had a design life of just ten years, but through a series of upgrades and modifications it served for more than twenty. But a series of mishaps over the years—some fatal—combined with ever-rising operational costs and a desire to modernize the country's nuclear deterrent persuaded the government that Titan II had at last outlived its usefulness. In 1981 the Reagan administration announced that Titan II would be retired.[32] Deactivation began in 1982 and was completed five years later in 1987.[33]

Today the U.S. strategic nuclear deterrent lies primarily with the Air Force's silo-launched Minuteman missiles, the Navy's submarine-launched Trident missiles, and to a lesser extent than during the Cold War, Air Force manned bombers.

There is, of course, a great deal more to the story of the ICBM and Titan II, but these few details should suffice in providing a helpful context for the facts and figures found in this book.

Those interested in a more thorough discussion of the history of ICBMs are referred to the sources listed in the bibliography.

▶ Titan II's storable propellant, fast response time and in-silo launch capability were a winning combination. Its enormous power could lift high-yield nuclear weapons.

1 1 http://www.grc.nasa.gov/WWW/K-12/TRC/Rockets/history_of_rockets.html
2 Heinz Gartmann, *The Men Behind the Space Rockets,* (David McKay Company, New York, 1956), 35-47
3 Emily McMurray, editor, *Notable Twentieth-Century Scientists Volume 2,* (Gale Research, Inc., 1995), 768-771
4 G. Harry Stein, *ICBM: The Making of the Weapon that Changed the World,* (Orion Books, 1991), 22-26
5 Gartmann, 136-159
6 McMurray *(Volume 4),* 2093-2096
7 Stein, 22-86
8 Ibid., 87-88
9 R. Cargill Hall, "Early U.S. Satellite Proposals," in Emme, ed., *History of Rocket Technology,* 68
10 Roger Launius and Dennis Jenkins, editors, *To Reach the High Frontier: A History of U.S. Launch Vehicles,* (University of Kentucky Press, 2002), 149
11 Robert L. Perry, "The Atlas, Thor, Titan, and Minuteman," in Emme, ed., *History of Rocket Technology,* 143
12 Launius and Jenkins, 158
13 Jacob Neufeld, *Ballistic Missiles of the United States Air Force,* 1945-1960, Office of Air Force History, 1989, 70
14 The V-2 suffered from the same problems. It was expensive, inaccurate, and could deliver less than a ton of conventional high explosives.
15 Richard Rhodes, *Dark Sun: The Making of the Hydrogen Bomb,* (Simon and Schuster, 1995), 482-512
16 Perry, 143-144
17 Stein, 161
18 Launius and Jenkins, 77-78
19 Ibid., 74
20 Craig and Jungerman, 79
21 Launius and Jenkins, 80-84
22 Ibid., 80
23 Snark, deployed at Presque Isle AFB in Maine, was intercontinental in range but not ballistic in flight.
24 Launius and Jenkins, 84-85
25 Stein, 199-200
26 Launius and Jenkins, 150-151
27 Ibid., 153-154
28 David Stumpf, *Titan II: A History of the Cold War Missile Program,* (University of Arkansas Press, 2000), 31
29 http://www.boeing.com/commercial/747family/pf/pf_classics.html
30 Stumpf, 74
31 Ibid., 130-141
32 Ibid., 253
33 Ibid., 130-141

USAF PHOTO, TMM ARCHIVES

▶ The configuration of the control center, access portal, blast lock and silo were exactly the same from site to site. Other topside features varied in location.

CHAPTER 2

CONSTRUCTION

At the time of Titan II's birth, U.S. relations with the U.S.S.R. were not very good. In 1956 Soviet premier Nikita Khrushchev made his infamous "we will bury you" remark. In 1957 the Soviets launched Sputnik, and for most of 1958, politicos were consumed with the "missile gap." In 1959 Vice President Richard Nixon had a run-in with Khrushchev in the famous "kitchen debate." In 1960 Khrushchev was banging his shoe on his desk at the United Nations. The failed U.S. Bay of Pigs invasion of Cuba made things even worse in 1961. Indeed, there was a clear sense of urgency in the construction of Titan II launch sites, an urgency made absolutely visceral by the Cuban Missile Crisis, which occurred in 1962, near the end of construction.

To save time, engineers used a scheme called concurrence. In most construction projects, engineers complete all the design and work out all the bugs, then hand the blueprints to the construction crews. In concurrence, design and construction happen at the same time. The blueprints are grabbed from the drawing boards almost before the ink is dry—the engineers are always just one step ahead of the construction workers.

There is a well-known engineering maxim that says you can have things fast, cheap, or right, but you can't have all three at once. For example, if you want something fast and cheap it won't be right. Or, if you want something cheap and right it won't be fast. In the case of the Titan II, the Air Force wanted it fast and right, which meant it wouldn't be cheap.

Concurrence saves time but is expensive because all the bugs get worked out with a jackhammer instead of an eraser. Every day contractors would ask "What do we tear out today that we installed yesterday?" There were hundreds of change orders resulting in thousands of modifications. Yet everyone understood that "the race must be won," and a multitude of problems not withstanding, concurrence worked. From the first shovel in the ground to the last crew to go on alert, the entire project—54 sites in three states—took just three years, almost to the day, to complete.[1] The Titan Missile Museum's complex, 571-7, took 29 months to build, and was put on alert July 15, 1963.[2]

Titan II launch sites were built in the vicinity of Little Rock AFB near Little Rock, Arkansas; McConnell AFB in Wichita, Kansas; and Davis-Monthan AFB in Tucson, Arizona. Each of these three bases was assigned 18 missiles (plus one spare), referred to as a Strategic Missile Wing (SMW). Each wing was subdivided into two groups of nine missiles known as Strategic Missile Squadrons (SMS).

Construction occurred in three phases.

Phase One

In Tucson, heavy construction began on December 9, 1960 and was completed at all 18 sites just a few days before Christmas in 1961. As the term "heavy construction" implies, Phase One included all the excavation and concrete work. At any particular complex, Phase One

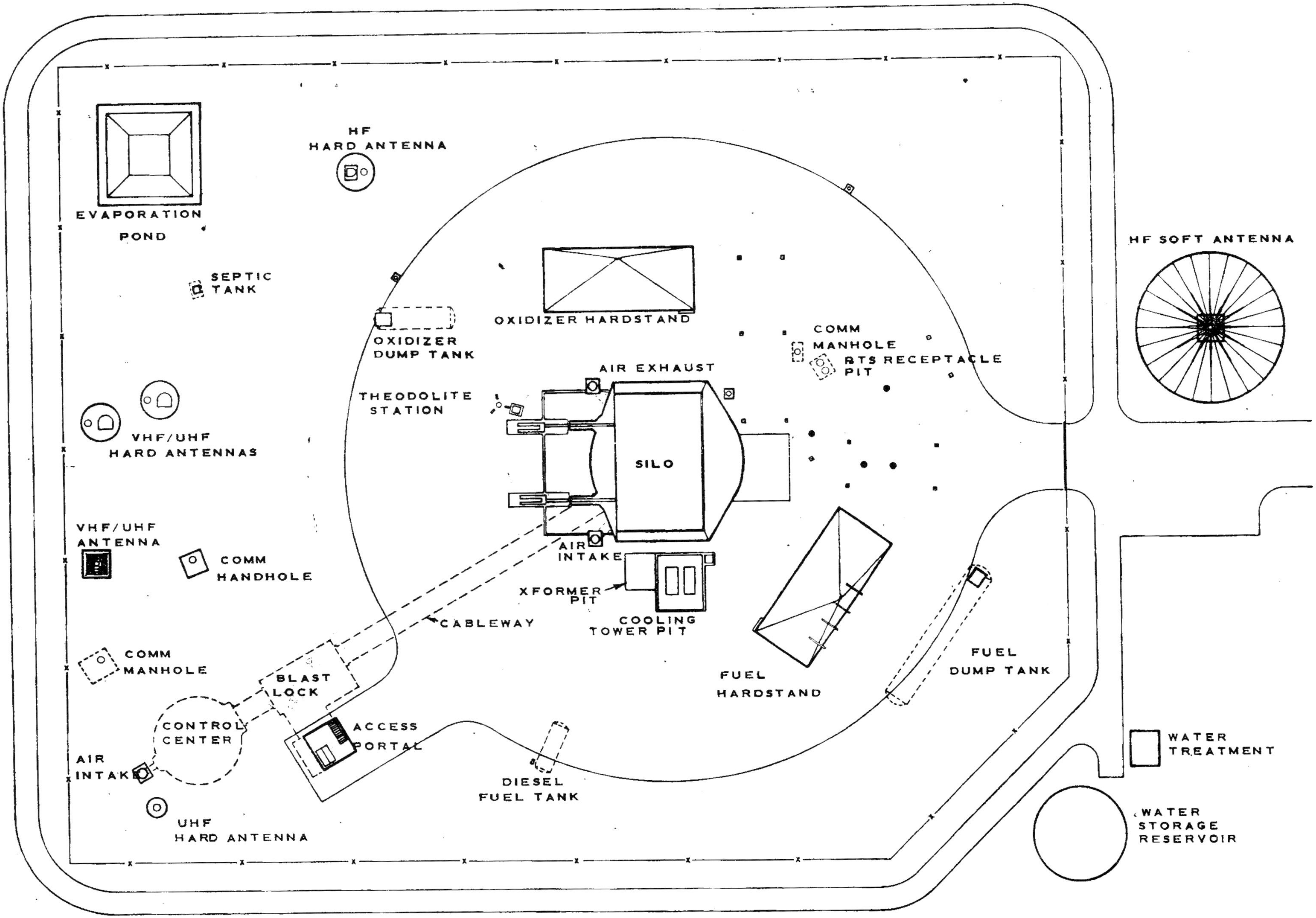
HF HARD ANTENNA
EVAPORATION POND
HF SOFT ANTENNA
SEPTIC TANK
OXIDIZER HARDSTAND
OXIDIZER DUMP TANK
COMM MANHOLE
RTS RECEPTACLE PIT
AIR EXHAUST
THEODOLITE STATION
VHF/UHF HARD ANTENNAS
SILO
VHF/UHF ANTENNA
AIR INTAKE
COMM HANDHOLE
XFORMER PIT
COOLING TOWER PIT
CABLEWAY
COMM MANHOLE
FUEL DUMP TANK
BLAST LOCK
FUEL HARDSTAND
CONTROL CENTER
ACCESS PORTAL
WATER TREATMENT
AIR INTAKE
DIESEL FUEL TANK
UHF HARD ANTENNA
WATER STORAGE RESERVOIR

▲ The "bathtub" with outline of the silo (right) and control center (left).

USAF PHOTO. TMM ARCHIVES.

construction took from seven to ten months, depending on the difficulties encountered in digging. Some "real property" was also installed during this phase. For example, the emergency generator was so large and heavy that it was installed upon completion of silo level 3 by lowering it into the unfinished silo with a crane. The rest of the silo was then built around it. Several pieces of equipment were installed in this way, and once installed they could not be removed, at least not in one piece, and in some cases, not under any circumstances.

Phase Two

Real Property Installed Equipment (RPIE) started at a given complex as soon as Phase One was completed. RPIE included all of the essential infrastructure: electric power systems; miles of electric wiring and conduits; lighting; plumbing; heating, ventilating and air conditioning systems; hydraulics; and so forth. Phase Two construction took about a year at each site.

Phase Three

Installation and Checkout started early in February 1962 and was complete at all Tucson sites by the middle of December. In Phase Three, contractors handed the keys to the Air Force. The Air Force then installed all of the electronics and communications systems in the launch control center (LCC), put in the hookups for the missile, and installed the missile itself. All Tucson sites were fully operational and on alert by October 1963. The 390th Strategic Missile Wing (SMW) in Tucson, composed of 18 launch facilities, was declared fully operational in December that same year.[3]

Hardening

The whole idea behind basing ICBMs in silos is to protect them from attack. The word

USAF PHOTO. TMM ARCHIVES.

heard most often in describing the construction of the launch sites is "hardened." Hardening refers to strengthening the complex so as to survive a nuclear attack, and can take several forms, the most obvious of which are massive amounts of concrete and steel. These offered protection against the powerful shockwaves that result from nuclear explosions. Concrete and steel also offered significant protection against two other nuclear effects. Nuclear weapons can generate both powerful electric and magnetic waves called electromagnetic pulses (EMP) and prompt radiation in the form of gamma rays, x-rays, and neutrons. EMP waves can burn out or disrupt sensitive electronic equipment such as missile guidance systems, while prompt radiation can be damaging to both equipment and humans. The effects of both can be blunted—but not entirely eliminated—with thick concrete walls and heavy metal shielding. Both were used extensively in Titan II launch sites.

There are two important points here: 1) any given amount of hardening will be effective only for a detonation beyond some specific distance from the silo, and 2) hardening costs money. Since no known structure is capable of withstanding a direct hit, the key question becomes "How much is enough?" Because there are so many variables in the equation, there is no simple answer.

Based on the presumed accuracy and yield of Soviet missiles, Titan II planners settled on hardening the sites to 300 pounds per square inch (psi) (21 kg/cm^2).[4] This meant that the site could withstand a shockwave generating 300 psi on the silo closure door and the ground covering the rest of the complex. In practical terms this translated (at least in theory) to withstanding a 10-megaton bomb exploding about 4700 feet (1.4 km) from the complex or a 20-megaton bomb at a distance of about 5800 feet (1.8 km).[5]

▲ With construction well underway, the blast lock and access portal (left) begin to take shape.

▲ Ring beams formed the outline of the silo wall, and supported the silo's exterior steel shell. Note bulldozer at bottom, and lift bucket.

▶ The silo was composed of 9-levels.

This does not necessarily mean that the missile would still be "good to go" after such a nearby strike, only that the complex itself would suffer no structural damage. Clearly, the shockwave traveling though the ground could "rattle" the silo so severely that, even with a spring suspension system to protect it, the missile might be rendered inoperative or sufficiently inaccurate to strike the intended target.[6]

As a hedge, engineers separated the silos by a minimum of 8 miles (14.5 km) so that one bomb could not destroy more than one silo.[7] That said, it is interesting to speculate that the Titan IIs may have been launched very early in a nuclear conflict and that their silos might have been empty by the time enemy bombs began to fall.

The vagaries of a nuclear exchange are many, and in the end engineers simply had to make the most educated calculations they could and hope for the best.

General Layout

The location of the access portal, blast lock, launch control center (LCC), silo, hardstands and related structures were identical at every complex. The LCC and silo were separated by a distance of about 250 feet (76 m), and the silos were oriented so that the doors opened from east to west. The LCC joined the silo through a *cableway*, a steel tunnel, which was oriented on a bearing of about 240 degrees relative to the silo: roughly west southwest.[8]

Placement of the many antenna systems varied from complex to complex but for practical reasons antennas tended to be clustered in the vicinity of the LCC.

Bathtub

Construction of the launch sites began with the excavation of what was referred to as the "bathtub." The bathtub was a large oblong cut into the ground, typically some 400 feet (120 m) long, 200 feet (60 m) wide, and about 40 feet (12 m) deep.[9] The actual size depended to some extent on the difficulty encountered in digging. For example, at hard rock sites the bathtub was often much smaller.[10] The major components of the complex—the access portal, blast lock, control center and silo—were all located within the bathtub. Once the bathtub was completed, further excavation of the silo and control center began.

Silo

The silo was excavated to a depth of about 155 feet (47 m) and was accomplished with a bulldozer essentially scraping its way down in ever-deepening circles, the dirt being removed with a lift-bucket.[11]

On the way down, steel ring-beams were installed every few feet to form the basic outline and structure of the silo wall. At the bottom of the silo, a ¼-inch (6 mm) thick steel plate floor was installed for electromagnetic pulse (EMP) shielding. Extending up from this floor, steel plates were attached to the ring beams to form a metal cylinder that extended all the way to the surface.[12]

▲ Another view of the silo ring beams. The two "wings" will permit installation of air shafts and shock delay pipes.

BOTH: USAF PHOTO. TMM ARCHIVES.

Between this outer steel shell and temporary movable internal wooden structures called slip forms, workers placed a massive framework

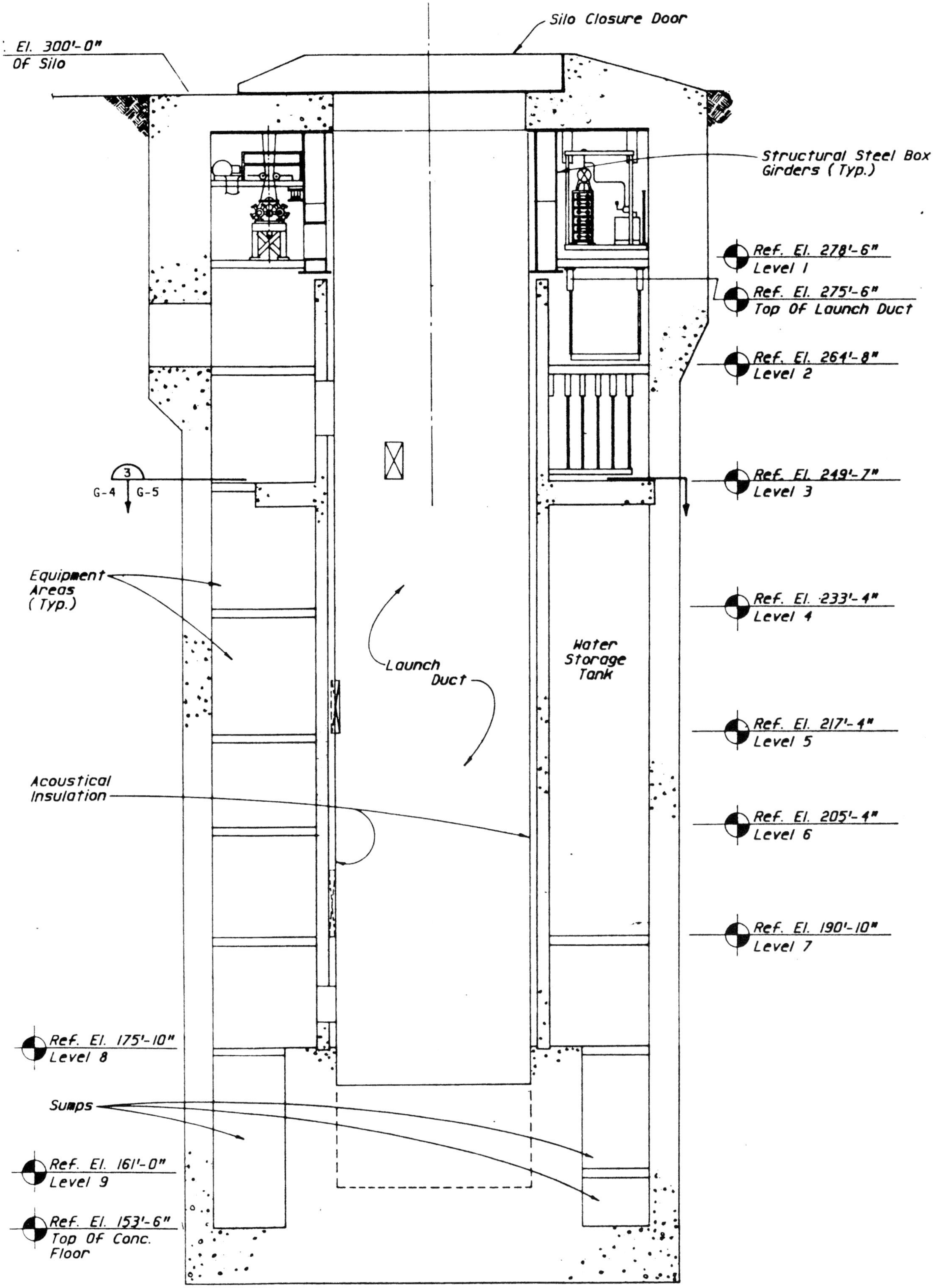

El. 300'-0"
Of Silo
Silo Closure Door
Structural Steel Box Girders (Typ.)
Ref. El. 278'-6"
Level 1
Ref. El. 275'-6"
Top Of Launch Duct
Ref. El. 264'-8"
Level 2
3
G-4
G-5
Ref. El. 249'-7"
Level 3
Equipment Areas (Typ.)
Ref. El. 233'-4"
Level 4
Launch Duct
Water Storage Tank
Ref. El. 217'-4"
Level 5
Acoustical Insulation
Ref. El. 205'-4"
Level 6
Ref. El. 190'-10"
Level 7
Ref. El. 175'-10"
Level 8
Sumps
Ref. El. 161'-0"
Level 9
Ref. El. 153'-6"
Top Of Conc. Floor

◀ The silo's funnel-shaped outer steel shell. The silo wall was four feet thick below the bottom of the bathtub, expanding to eight feet thick above. Note blast lock with opening for the long cableway.

▲ A huge steel "box girder" supported the silo door and "head works." Note shock delay pipes protruding from ground at lower right.

▲ Looking down the silo before construction of the launch duct. Note flame deflector.

ALL ON THIS PAGE: USAF PHOTOS, TMM ARCHIVES.

of nearly 850 tons (770 metric tons) of steel reinforcing bars, often called "rebar." Much of this rebar was as large as 2¼ inches (6 cm) in diameter.

With the rebar in place, concrete was poured between the outer steel cylinder and the inner slip forms creating a steel-reinforced wall four feet (1.2 m) thick. As the concrete behind the slip forms solidified the forms were raised (slipped up) either by a crane or with hydraulic jacks, and pouring continued. By pouring at a rate of about one to two feet (30 to 60 cm) per hour it was possible to pour continuously.[13] Concrete was mixed in batching plants located on site to ensure an uninterrupted supply. Water for making the concrete was often purchased from nearby ranches and supplied via pipe laid on the ground, sometimes for distances as great as four miles (6.5 km).[14] On average, a silo pour took about three days to complete.[15]

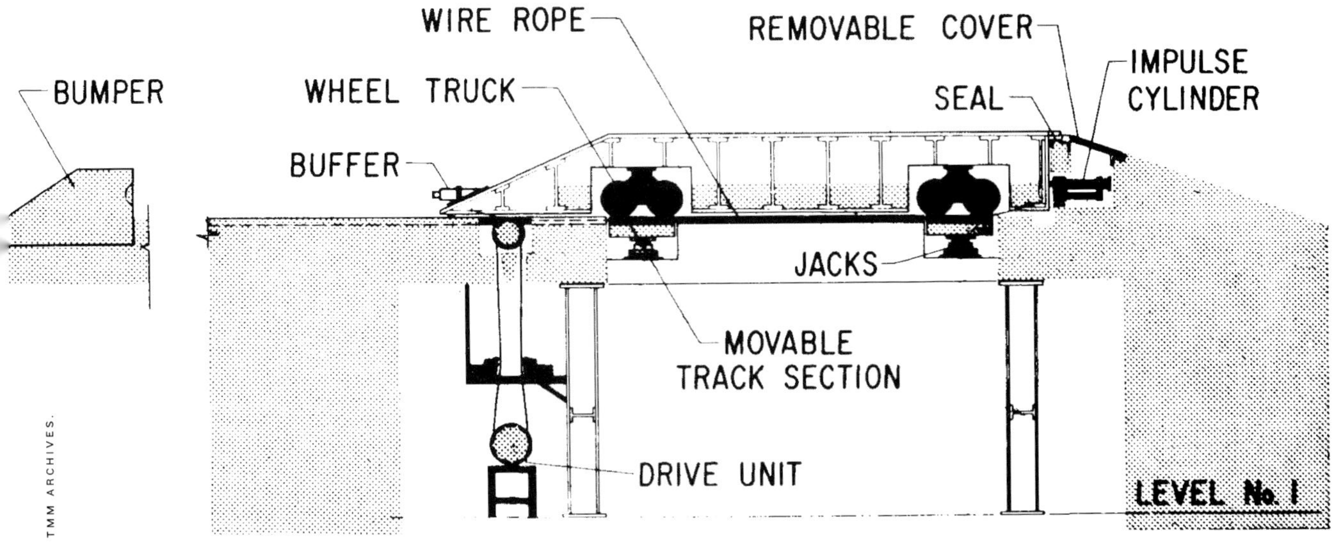

◀ The silo door was a marvel of engineering and could be fully opened in less than 20 seconds.

TMM ARCHIVES.

The top 30 feet (9 m) of the silo wall was increased to a thickness of eight feet (2.4 m). This additional girth helped to support not only the weight of the 725-ton (657-metric ton) silo door (later modifications brought the weight of the door to 760 tons), but also the added load that would be placed upon the door and silo by the shockwave of a nuclear detonation. Its mass also helped to damp shockwaves and permitted the silo to flex slightly in response to shock.[16]

With the completion of the silo wall, construction began on the launch duct, a smaller silo, 26.5 feet (8 m) in diameter, standing within the larger silo. The wall of the launch duct was a kind of concrete sandwich: 18 inches (46 cm) of concrete between two half-inch (1 cm) steel plates.

At the bottom of the launch duct was a large concrete wedge, referred to almost universally as the "W" because of its shape in profile. The W served as a flame deflector and directed the missile's engine exhaust through two vertical shafts to vents on the surface. At 571-7 these vents have been covered with concrete for the safety of visitors.

From the top of the silo to the bottom of the sumps was 147 feet (45 m). To the bottom of the foundation added another 7 feet (2 m).[17]

Silo Door

Of all of the structures built for the Titan II launch sites, the silo door was arguably the single most complex component. The door was 43 feet (12.5 m) long, 64 feet (19.5 m) wide, and 5 feet (1.5 m) thick at the center. The top and bottom surfaces, and many of the internal

◀ Meeting the specified precision with steel as thick as that used in the silo door had never before been attempted. Much trial and error was required to find a technique that worked.

USAF PHOTO. TMM ARCHIVES.

▶ The door rode on 16 wheels, in four sets of four. To open, hydraulic jacks lifted the wheel-truck rail-bridges level with the main rails.

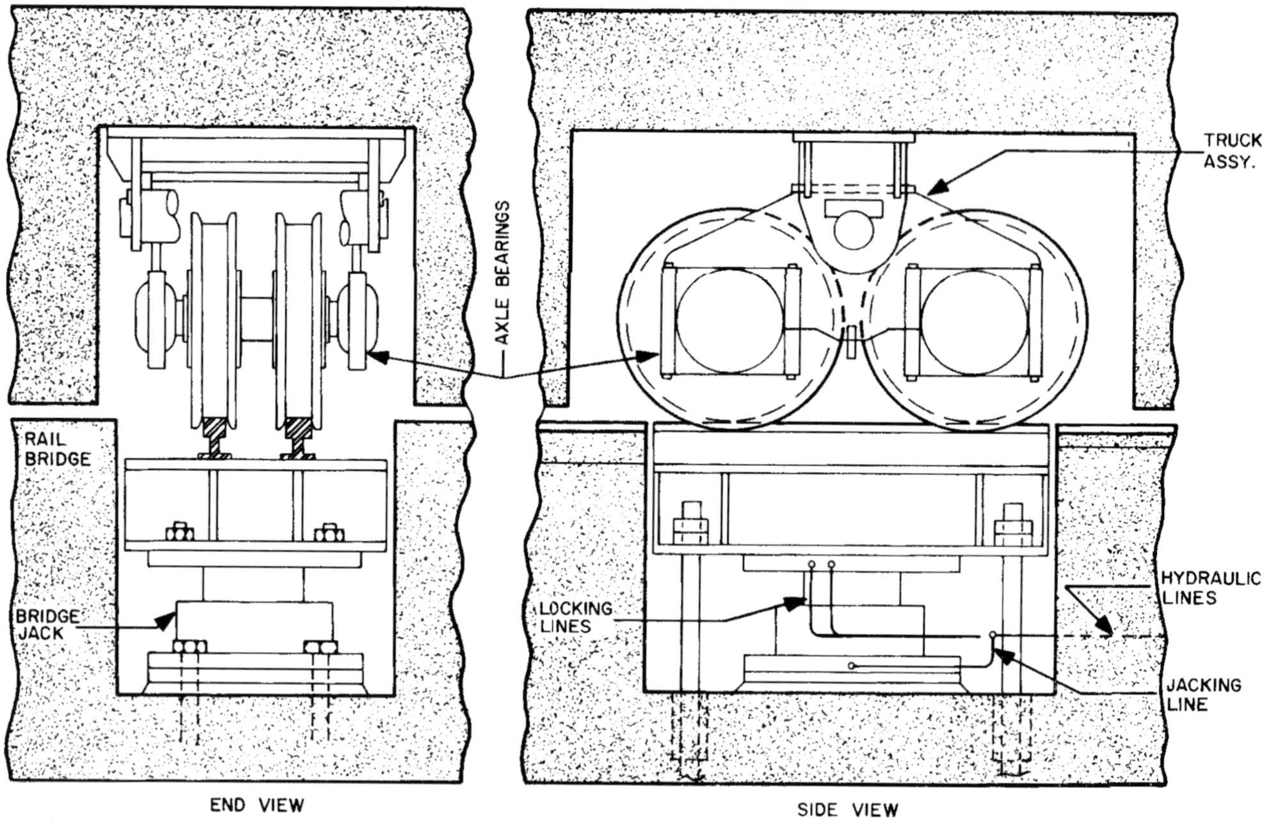

TMM ARCHIVES.

▼ The silo door wheels ran on tandem rails. Note rail-bridge at lower left, and exhaust vent at far left.

◀ "Dog ears" on the north and south sides of the door (left and right in the photo) were added two years after completion to provide more radiation shielding to compensate for the decreased thickness of the sides of the door.

USAF PHOTOS, TMM ARCHIVES.

support members were fabricated from steel slabs 2.5 inches (6 cm) thick. The structure was designed to withstand an overpressure (the pressure applied to the door by the shockwave of a nuclear detonation) of 300 psi (21 kg/cm^2).

Silo door sections were prefabricated locally and transported to the work sites by rail or truck for final assembly. The internal structure of the door is often described as being shaped like an egg carton and was filled with about 15 inches (38 cm) of concrete. More than 37 tons (33.5 metric tons) of welding rods were used to fabricate each door.[18]

The most significant problem in building the door was welding its massive steel plates. The entire bottom surface of the door was

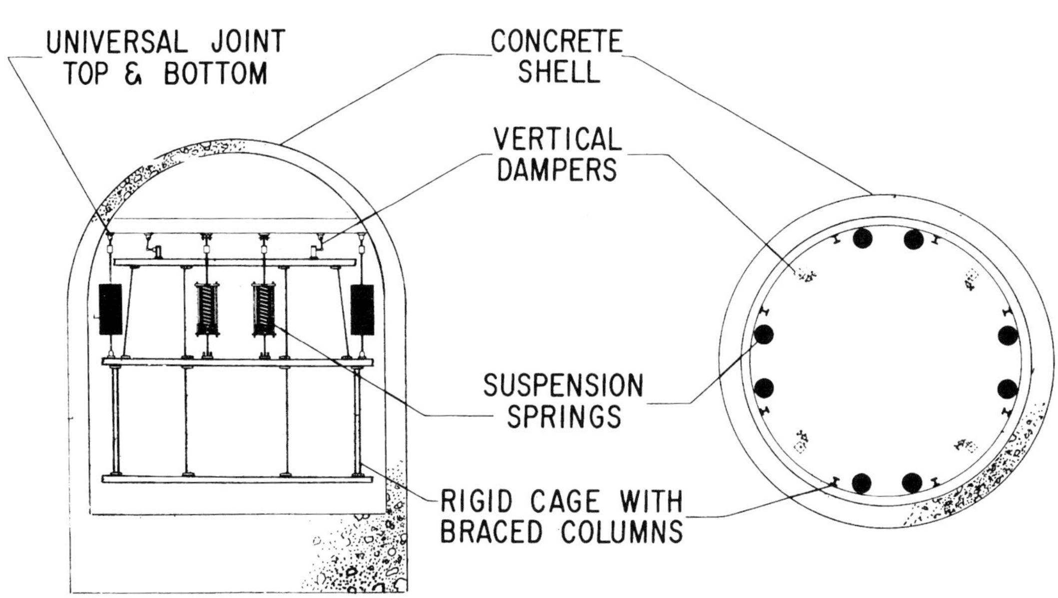

◀ Schematic drawing of the control center and spring suspension system.

TMM ARCHIVES

SHOCK ISOLATION SYSTEM CONTROL CENTER

specified to be flat within one sixteenth of an inch (1.5 mm). Heat from welding caused the plates to warp, pushing them out of this specification. Nothing seemed to help. Experts from government and private industry were consulted to help find a solution. In a desperate attempt to solve the problem contractors even tried refrigerating the metal and welding only at night. The ultimate solution was fairly simple: a change in design of the door's beryllium seals allowed the flatness specification to be relaxed, and pre-cambering the plates helped reduce the amount of distortion as the welds cooled.[19]

There is some disagreement regarding the weight of the door. Obviously it was never possible to put the door on a scale. Its weight was extrapolated from the mass of its constituent parts. Still, various official documents provide different numbers—anywhere from 699 to 749 tons (634 to 679 metric tons). The issue is further complicated by a later modification (discussed shortly) that increased the weight of the door. This guide will use as its starting point the value stated in "Final Report, Silo Door Test Program," drafted by the Ralph M. Parsons Company, dated June 1962. This report lists the average weight of the doors as 725 tons (657 metric tons).

The door rolled open and closed on sixteen 30-inch (76 cm) wheels (8 were double-flanged) configured in four sets of four, riding on a pair of tandem railroad tracks.[20,21,22,23]

The door was opened and closed with a hydraulic winch powered by high-pressure nitrogen gas. This meant the door could be opened without commercial or back-up generator electric power.[24] Emergency battery power to operate a system of valves was the only electricity required.

When closed, the door was locked in place with its wheel trucks resting on short lengths of track called rail bridges. These sections were about three inches (8 cm) lower than the main track. To open the door the locks disengaged and hydraulic jacks lifted the rail bridges—and hence the entire door—level with the main tracks. Steel cables attached to the winch then pulled the door from east to west a distance of about 30 feet (9 meters) in about 20 seconds. The actual specification called for an opening time of not to exceed 21 seconds. In one Air

▲ Workers install rebar in the 8-foot thick foundation of the control center. These bars are among the heaviest used on the complex.

USAF PHOTO, TMM ARCHIVES

◀ The two vaults extending from the control center walls accommodate radio and electrical cables and the air intake shaft. The large "donut" connects to the air intake shaft. An identical donut below (not visible) connects the escape hatch.

◀ The control center dome with interior concrete forms in place, prior to installation of reinforcing steel. Note workers for scale.

USAF PHOTOS, TMM ARCHIVES.

LEFT AND BOTTOM: USAF PHOTOS. TMM ARCHIVES.

▲ With reinforcing steel in place, the dome is ready for concrete pour.

▲ Three examples of reinforcing bar typical of the size used to build the underground structures (top). Rebar that might be used in residential construction (bottom).

Force video the door can be seen to open in 15 seconds flat.[25,26] Large hydraulic piston buffers were installed on the door to prevent it from slamming against the bumpers on the west side.

Specifications called for the door to be able to operate with six inches (15 cm) of debris calculated to weigh 50 tons (45 metric tons) on its top surface, while simultaneously pushing through six inches of debris on the ground with its plow-like leading edge.[27] This specification, however, was set before neutron-shielding "dog ears" were added to the north and south faces of the door as part of a modification begun in 1965. The dog ears provided extra protection from neutrons, compensating for the decreased thickness of the sides of the silo closure door.[28] Calculations by the author place the weight of the dog ears at approximately 18 tons each. If accurate, this would have brought the total

◀ A steel shell protected the entire complex. In this image the blast lock shell is being installed in preparation for pouring its concrete walls.

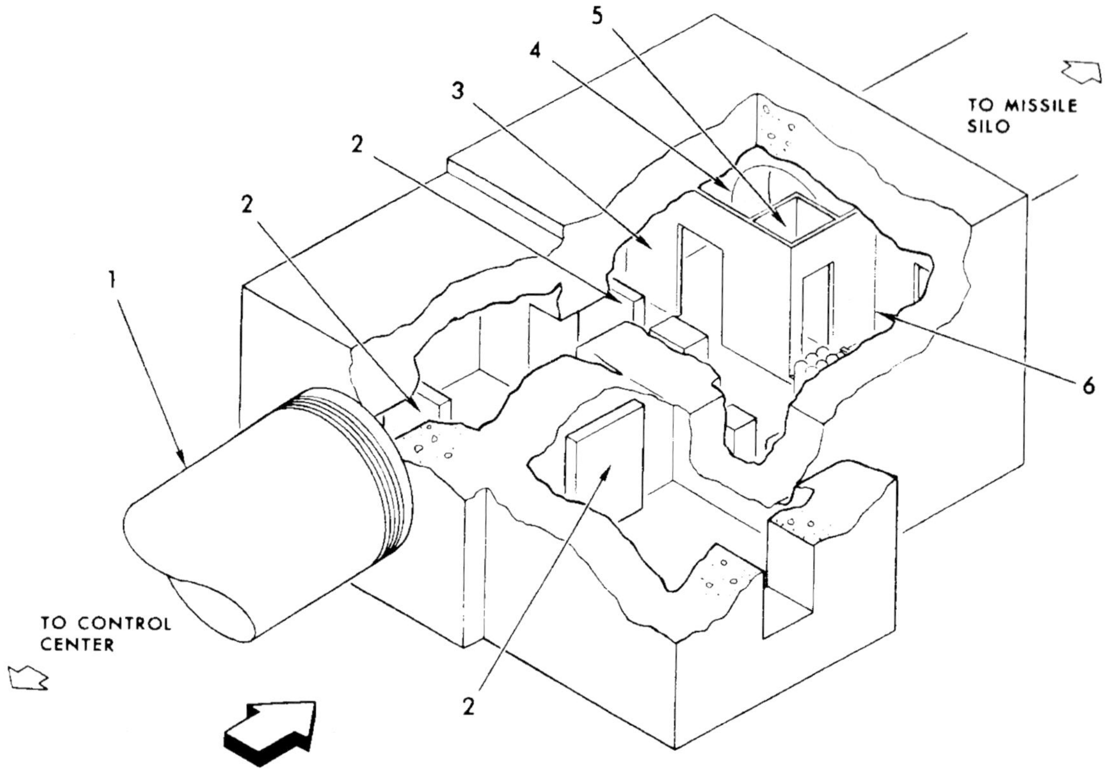

BLAST LOCK

1. CABLEWAY
2. BLAST DOOR
3. REJUVENATION AREA
4. DECONTAMINATION AREA
5. DECONTAMINATION SHOWER
6. DRESSING ROOM
7. BLAST LOCK ENTRANCE
8. PERSONNEL ENTRANCE
9. FREIGHT ELEVATOR ENTRANCE
10. PERSONNEL STAIRS
11. FREIGHT ENTRANCE

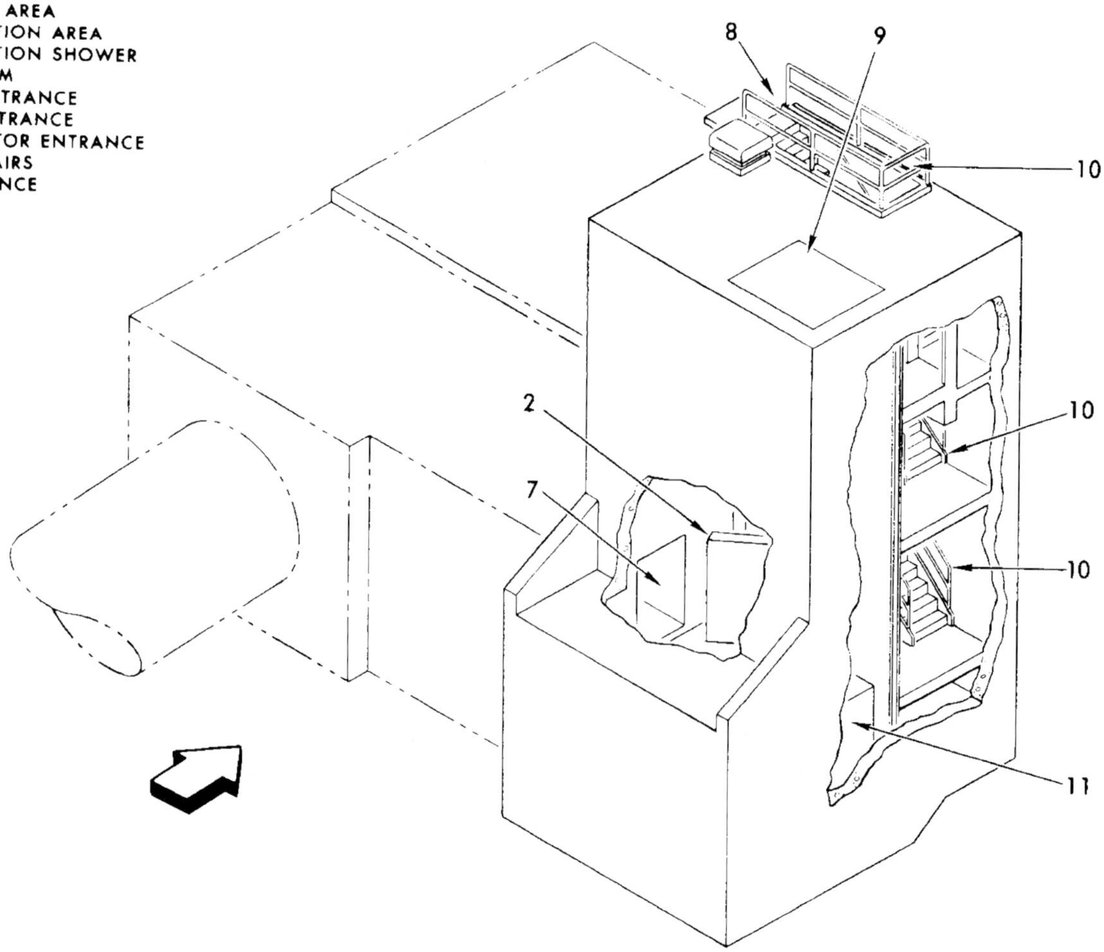

ACCESS PORTAL

USAF PHOTOS, TMM ARCHIVES.

▲ The access portal was not protected with a steel shell. It was regarded as "soft" and expendable. When construction is complete, the top of the access portal will be at ground level.

▲ The nearly completed blast lock structure. The form used to make the round opening for the cableway tunnel has been extracted and sits nearby.

weight of the door to approximately 760 tons (700 metric tons). How this added weight affected the door's ability to open with debris on top apparently was never determined.

Launch Control Center

As work on the silo progressed, the blast lock and launch control center (LCC) began to take shape as well.

The LCC consisted of four basic components: a heavy concrete foundation, a cylindrical midsection, a domed roof, and an independent internal structure. The top of the dome was about 18 inches (46 cm) thick and lay just nine feet (2.7 m) below the surface of the ground. The cylindrical midsection had walls about two feet (60 cm) thick, and sat on a foundation eight feet (2.5 m) thick, reinforced with rebar 2 ¼ inches (6 cm) in diameter. The inside of the LCC was 37 feet (11 m) in diameter and 42 feet (13 m) tall. Like the silo, the LCC was steel plated for protection from EMP.[29]

What made the LCC so remarkable was its fourth component, the independent internal structure—three floors contained within the concrete shell but not attached to it. All three levels were connected together and suspended from eight gigantic spring assemblies around the perimeter of the facility. Entering the LCC, one crossed a small bridge that prevented falling into a gap about a foot (30 cm) wide between the floor and the wall. This gap was referred to as the rattle space and permitted the entire three-level structure to bounce up and down and swing from side to side in response to a shockwave.

◄ The blast lock and control center are complete. A short tunnel will connect the two structures.

Blast Lock and Access Portal

The blast lock was a separate structure connecting the LCC to the silo, and both of these to the outside world via the access portal. The blast lock provided physical security by preventing unauthorized persons from gaining access to the complex. At the same time, it allowed easy access for personnel while simultaneously offering protection from the heat, radiation, and powerful blast waves created by a nuclear detonation. These functions were accomplished by four massive steel doors: one at each end of two heavily fortified concrete rooms. One pair of doors connected the blast lock to the access portal (and thus to the outside world) and served primarily to protect the

◀ The site's four massive blast doors. The blast lock will be built around them. Blast door 6 is at left, door 7 at center, and door 9 at right. Door 8 is not visible.

complex from a nuclear blast. Another pair of doors connected the control center to the silo and served primarily to protect the crew from their own missile.

The basic idea of the blast lock was the same as an air lock in a spaceship or a submarine. Only one door of a pair could be opened at a time. Electric interlocks prevented both doors from being opened together. Since at least one door was always closed, a blast wave could never fully penetrate the complex.

In total, some 2255 tons (2000 metric tons) of steel and about 8400 cubic yards (6400 cubic meters) of concrete were used in each of the 54 Titan II launch sites.[30,31]

The price tag in 1960 dollars was about $12.6 million per complex, including about $2.5 million for the missile itself—a total of almost a quarter of a billion dollars for the 18 sites in Tucson alone.[32] Adjusted for inflation this translates to about $1.8 billion in 2011 dollars, or about $100 million per complex, including the missile. Construction was costly in human terms as well. Four workers lost their lives building the sites around Tucson. Many more were injured, some seriously.[33] The project required the combined effort of more than 5000 workers in each of the three states involved. The economic infusion for the surrounding communities was huge—and it didn't stop once construction ended. Millions of dollars were required annually to operate and maintain the sites and their missiles over the next twenty-plus years.

▼ Finishing work on the silo headworks. The silo door is complete and in place. This photo looks west. The door opens from east to west.

USAF PHOTOS, TMM ARCHIVES.

USAF PHOTO, TMM ARCHIVES

▲ **Backfill operations in progress. When complete, the top of the control center will be about 9 feet below ground level.**

1 Stumpf, 130-141
2 Ibid., 130-131
3 Ibid., 135
4 Ibid., 99-102
5 Samual Glasstone, Editor, *The Effects of Nuclear Weapons,* Department of the Army Pamphlet No. 39-3, 1962, 121-148
6 Kosta Tsipis, *Arsenal: Understanding Weapons in the Nuclear Age, (Simon and Schuster*, 1983),136-139
7 Strategic Air Command, *Master Plan Regional Map*, Davis-Monthan Air Force Base, May 1962, Tab (sheet) B-1, TMM archives.
8 TO.21M-LGM25C-1, 1-6. Note: All references to TO.21M-LGM25C-1 and 00-ALCP-144-4 are to page numbers, not paragraph numbers.
9 TMM archives. Description on construction photo.
10 Stumpf, 111
11 TMM archives. Photos of construction.
12 TMM archives. Photos of construction.
13 Stumpf, 115
14 Harry Christman, *The Construction of the Intercontinental Ballistic Missiles at Davis-Monthan AFB, Tucson AZ (The Fluor Report), unpublished, 1999, 14*
15 Stumpf, 155
16 Ralph M. Parsons Company, Final Design Calculations, Structural, Volume 1 Section 4D, *Launch Silo Launch Duct: Relative Movement Between Duct and Silo,* sheets 1 through 9, October 1961
17 SM-68B Technical Facilities Standard Plans, sheet 105/66 (operational as-built blueprint), TMM archives
18 Stumpf, 120-124
19 Christman, 11-13
20 Ralph M.Parsons Company, *Final Report Silo Door System Test Program,* WS 107 A-2 Titan II, 1962, 4-9, TMM archives
21 Stumpf, 101
22 *Weapon System Familiarization, Course OZR1821F/3121F*, Technical Training, Shepard AFB, January 1965, 16
23 TMM archival photograph, number 31876
24 TO.21M-LGM25C-1, 1-8, 1-9
25 Parsons, *Final Report,* 1-3, 2-5, 4-1, 4-2
26 Titan Missile Museum archives. Door travel time as measured in Air Force training video "Launch Verification."
27 Parsons, *Final report,* 1-4, 2-4
28 Operation Yard Fence, 1965
29 SM-68B Technical Facilities Standard Plans, sheet 105/78 (operational as-built blueprint), TMM archives
30 Stumpf, 112, 117
31 Christman, 14
32 Stumpf, 146
33 U.S. Army Corp of Engineers, *History of the Army Corp of Engineers Ballistic Missile Construction Office Titan II Facilities Davis-Monthan Air Force Base Tucson Arizona*, 1962, 7.9-7.12

► In most cases this sign was located on the main gate of the complex. While the warning seems stern enough, some versions added "USE OF DEADLY FORCE AUTHORIZED," to make sure people got the message.

CHAPTER 3

TOPSIDE

Titan II launch sites were located on about ten acres (4 ha) of land, but only the center three acres (1.2 ha) or so were used to house the silo, control center, and related facilities and equipment. The remainder of the land was mostly buffer. A seven-foot (2 m) tall fence topped with barbed wire served as a perimeter and provided security for the central portion of the complex. The area inside the fence was maintained with a gravel surface and had none of the pathways seen at 571-7 today. These pathways were installed for the safety and convenience of visitors.

It is important to understand that none of the equipment seen topside at the Titan Missile Museum was present on a typical day. This includes all the trucks, cars and trailers, and all of the equipment associated with the loading and unloading of propellant. The special equipment necessary for propellant handling was brought to the complex only as required, and on a normal day someone looking through the fence would see only the silo door and the various poles and antennas situated about the surface. There would be no sign of crews or maintenance teams.

▼ Concrete hardstands served as parking surfaces for propellant tank trucks and holding trailers. A "conditioning trailer" sits at left.

◄ Propellant was loaded into the missile by gravity via stainless steel piping. The larger of the two pipes in the foreground is for stage one, the smaller one for stage two.

Propellant Hardstands

On the north and east sides of the silo were two large rectangular concrete pads called

▲ Contractors put the finishing touches on a complex near Wichita, Kansas.

▶ Water spray nozzles on the fuel hardstand for dilution of spills or fire suppression. No such nozzles were present on the oxidizer hardstand.

USAF PHOTO, TMM ARCHIVES

hardstands. These served as parking areas for the tank trucks and holding trailers used to load and unload missile propellant. Fuel was loaded from the east hardstand and oxidizer from the north. The most obvious difference between the two was the fuel hardstand's water spray nozzles. These nozzles were used for fire suppression and to wash down equipment if a fuel spill occurred. In the early days of the program, the fuel and water runoff was directed into a drain that emptied either at grade level just east of the hardstand or into underground dump tanks (discussed shortly).

The oxidizer hardstand had no water spray system. Mixing water with oxidizer would result in a considerable volume of poisonous vapor characterized by a large red-orange cloud, referred to informally by many Air Force personnel as a "BFRC" (Big [-expletive-] Red Cloud).

Security Equipment

No guards were stationed topside, and no TV cameras provided the crew with a view of the complex. The perimeter fence kept out most animals and all but the most determined intruders (though intruders were never a significant problem). Outside the fence there

▶ A plume of highly toxic nitrogen tetroxide vapor wafts away from an oxidizer transport trailer during a PTS operation. A certain amount of venting was unavoidable.

USAF PHOTO, TMM ARCHIVES.

▶ The "tipsie" security system projected a motion-sensing radar beam around the silo and over the control center air intake shaft.

was no form of security. Inside the fence security was maintained with the AN/TPS-39 radar surveillance security system.[1] Because of the "TPS" in the designation, virtually everyone took it to mean "territorial protection system." In reality, TPS stood for something quite different. According to the Department of Defense Standard Practice, Joint Electronics Type Designation System (MIL-STD-196E) the T stood for transportable (nevermind that it wasn't), the P for radar (R had already been taken by "radio"), and the S for search. Still, territorial protection system made perfect sense to everybody who didn't know otherwise. Air Force personnel reduced TPS to "tipsie" and the nickname quickly became common usage.

Pairs of tipsie units were located at each corner of the silo, and also over the LCC air intake shaft. Radar beams sent between these units created an electronic motion sensor. If a moving object disturbed any of the beams, the system set off an alarm in the control center. The crew on alert would not come topside to investigate. They would call the wing command post (WCP), and the WCP would dispatch the security police. The launch complexes were associated into groups of three sites called sectors. The sites were located such that security police were rarely more than 20 minutes away from any complex in a given sector.

Twenty minutes might sound like a long response time, but there was usually no real hurry. While the Air Force was not happy about people snooping around inside the fence, it also realized there wasn't much a would-be saboteur could do on the surface to interfere with a launch.[2] Over the history of the program there were very few instances of genuine intrusions topside, and none at all underground. In the early years virtually all trips of the tipsie system were false alarms caused by large birds, heavy rain or other anomalous sources.[3] The system was eventually desensitized to reduce false alarms.[4]

The tipsies had their own battery backup

▲▲ Silo door being moved from fabrication site to silo.

▲Bumpers and shock-absorbing pistons brought the door to a stop. The large concrete cubes were installed by the museum to assure the Soviets that the door could no longer be fully opened.

system and could continue to operate for two hours in the event of a complete power failure.[5] If the batteries went dead, guards would be posted.

Silo Door

The silo door weighed an estimated 760 tons (700 metric tons) and could be fully opened in about 20 seconds. At 571-7 the silo door is exactly halfway open. Large exhaust ducts on the north and south sides of the silo, which vented the missile's exhaust during launch (see photo page on 89) have been covered with concrete, as have the rails on which the silo door rode. Part of the rails can still be seen where they emerge from under the west side of the door. When 571-7 became a museum large concrete blocks were installed on the rails to ensure that the door could no longer be opened all the way. The blocks were added as part of an agreement with the Soviet Union to help establish the site as a "non-counter," meaning that it would not be included in the number of active missile launchers permitted to the United States by treaty.

A complete discussion of the silo door can be found in Chapter 2.

▲ The notoriously hard-to-photograph HF discage antenna was made by Collins Radio and covered from 3 to 30 MHz.

Communications Antennas

A Titan II launch crew had a lot of tasks to keep them busy while working underground, but one of their primary occupations was to listen for the order to launch their missile. It follows then, that staying in touch with higher headquarters was of primary importance. To

◀ A hawk scouts for lunch atop the discage antenna.

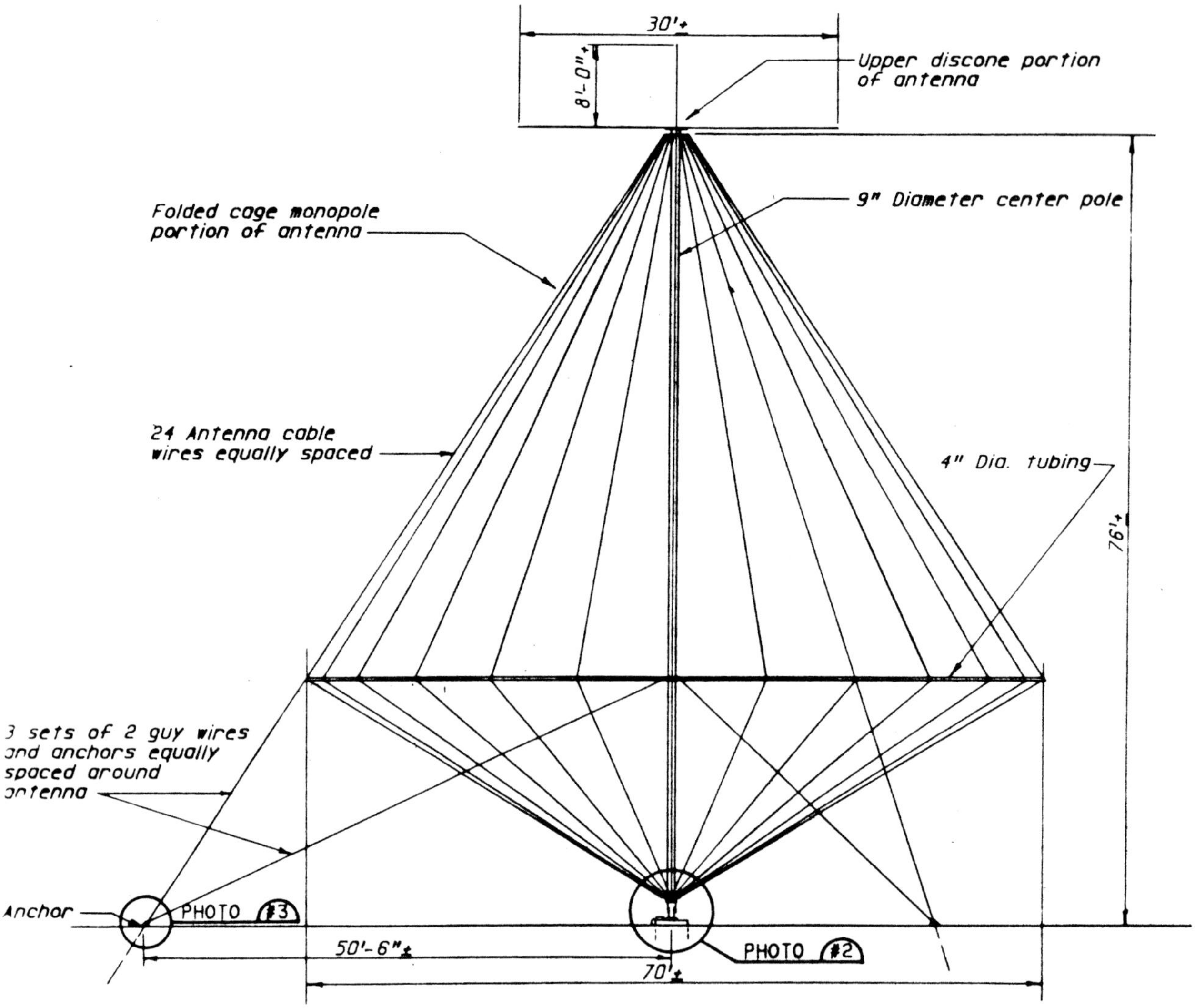

ELEVATION
HF DISCAGE ANTENNA

TMM ARCHIVES

▶ The hardened HF antenna, also made by Collins, telescoped out from its own small silo.

make sure the message got through, each launch complex had several communication systems, as evidenced by the numerous radio antennas seen topside.

High Frequency Soft Antenna

The most obvious antenna was the 76-foot (23 m) discage.[6] Because of its size the discage was located outside the main security fence. At 571-7 this antenna is located near the museum's parking lot. The discage takes its name from the disk-like configuration of metal poles at the very top, and the cage-like main body of the antenna. It is actually two antennas. It could function as a discone, operating from 6 to 30 megahertz (MHz), or as a folded-cage monopole, operating between 3 and 6 MHz. This portion of the radio spectrum (3-30 MHz) is officially referred to as the high frequency (HF) band, but is more commonly known as shortwave radio. The discage was a soft antenna, meaning that it was not expected

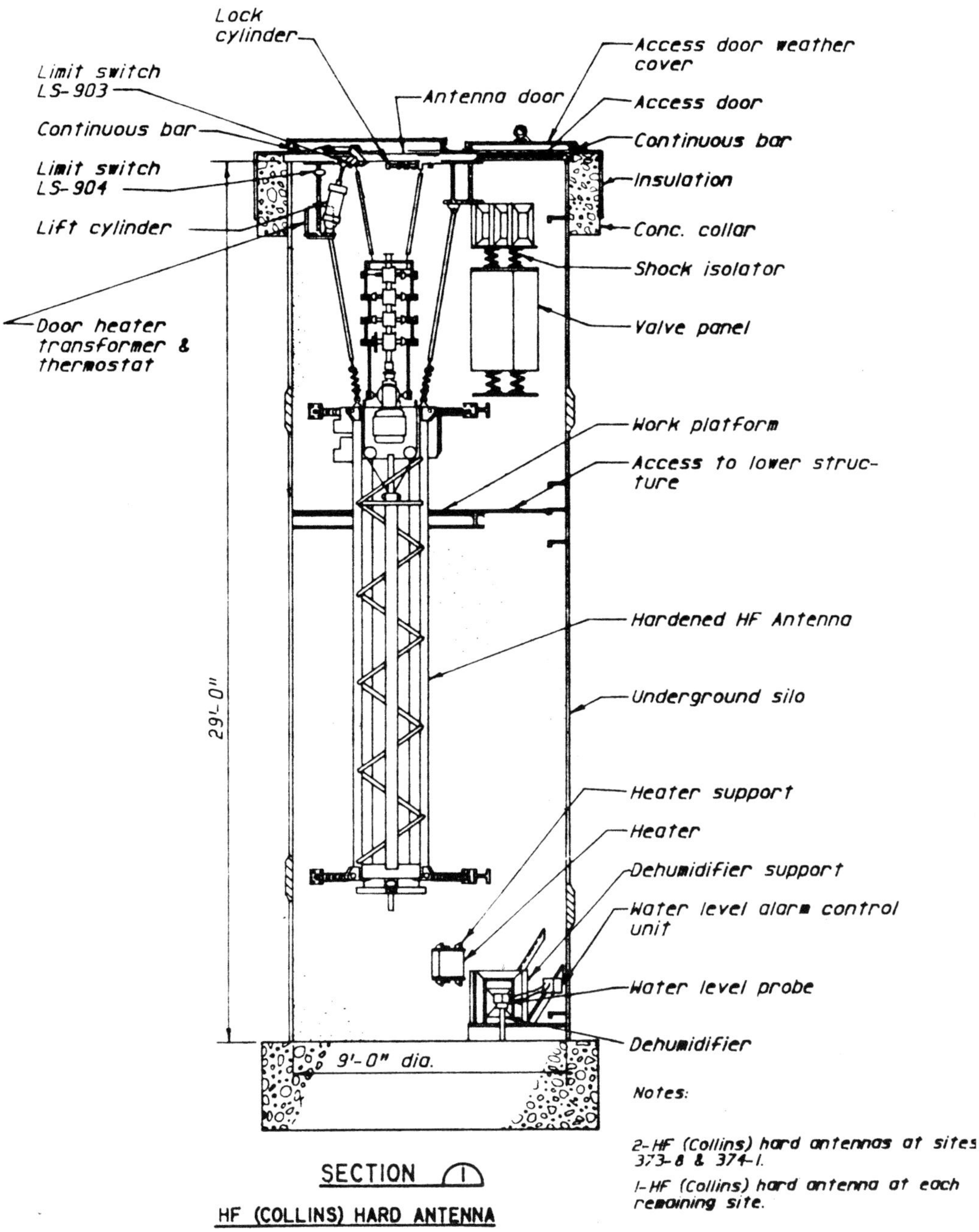

TMM ARCHIVES

to survive a nearby nuclear detonation. The antenna at the Titan Missile Museum still functions and is often used by visiting amateur radio operators.

High Frequency Hard Antenna

To maintain HF communications capability if the discage were destroyed, the discage was backed up by a hardened HF antenna normally stowed safely underground in its own small silo. Located near the northwest corner of the complex at 571-7, this telescoping antenna could be raised and lowered by controls in the LCC.

◀ IRCS silo top and antenna hatch. The round cover is for maintenance access.

▲ The intercomplex radio communication system (IRCS) antennas, made by General Electric. The soft antenna is at left, with one of two hardened backups at right.

The antenna covered the same frequency bands as the discage but its operating frequency was determined by the height to which it was raised. It could be raised from 7½ feet (2.5 m) to as high as 117 feet (36 meters).[7] The antenna silo was about 30 feet (9 meters) deep. The antenna door was opened with high-pressure nitrogen gas, but the antenna itself was raised and lowered with electric motors and required commercial or generator electric power to function.

Intercomplex Radio Communication System Antennas

The intercomplex radio communication system (IRCS) was equipped with soft and hard antennas and was the most vital radio system on the complex: it was the primary means of receiving the order to launch. The IRCS was designed for maximum survivability and employed two backup antennas.

The IRCS antenna was a white pole atop a short tower, and at 571-7 can be seen just outside the west perimeter fence. This antenna was designed for frequency diversity, meaning that it could operate on several different frequencies simultaneously. The system carried both voice and data signals and was used to provide communications with the wing command post (WCP), two alternate command post launch sites (ACP)[8] and the other 16 basic launch complexes (BLX), and could relay messages from Strategic Air Command (SAC) headquarters. The IRCS carried routine message traffic as well as emergency action messages (EAMs). The IRCS would carry the order to launch via the primary alerting system (PAS) channel—essentially a direct link to SAC headquarters near Omaha, Nebraska.[9]

▲ The Mt. Lemmon communications site. Note the now-idle IRCS antenna at left. The other antennas are still active.

Because the Tucson sites were located in mountainous terrain that could potentially interfere with communications, the IRCS employed a mountaintop radio relay system. Hikers on the summit of Mount Lemmon can see an identical antenna inside the fence of an operational Air Force radio site. While the Mount Lemmon IRCS antenna is still in place, it is no longer in use. Such a relay system was not used in Arkansas or Kansas where the terrain was relatively flat.

Like the discage, the IRCS antenna was soft, making it unlikely to survive an attack. Maintaining the functionality of the IRCS was so important that the system had not one, but two backup antennas. Had the soft antenna been destroyed, two identical "pop-up" antennas could have been deployed from their own hardened silos and could be operational in just 50 seconds.[10] Only one of these pop-ups would be deployed. The other would be used only if the first was subsequently destroyed. The pop-up antennas and antenna silo lids were operated by high-pressure nitrogen gas. Originally configured to rise automatically if the soft antenna tilted by more than about 10 degrees, the system was changed to manual operation, requiring a crewmember to press

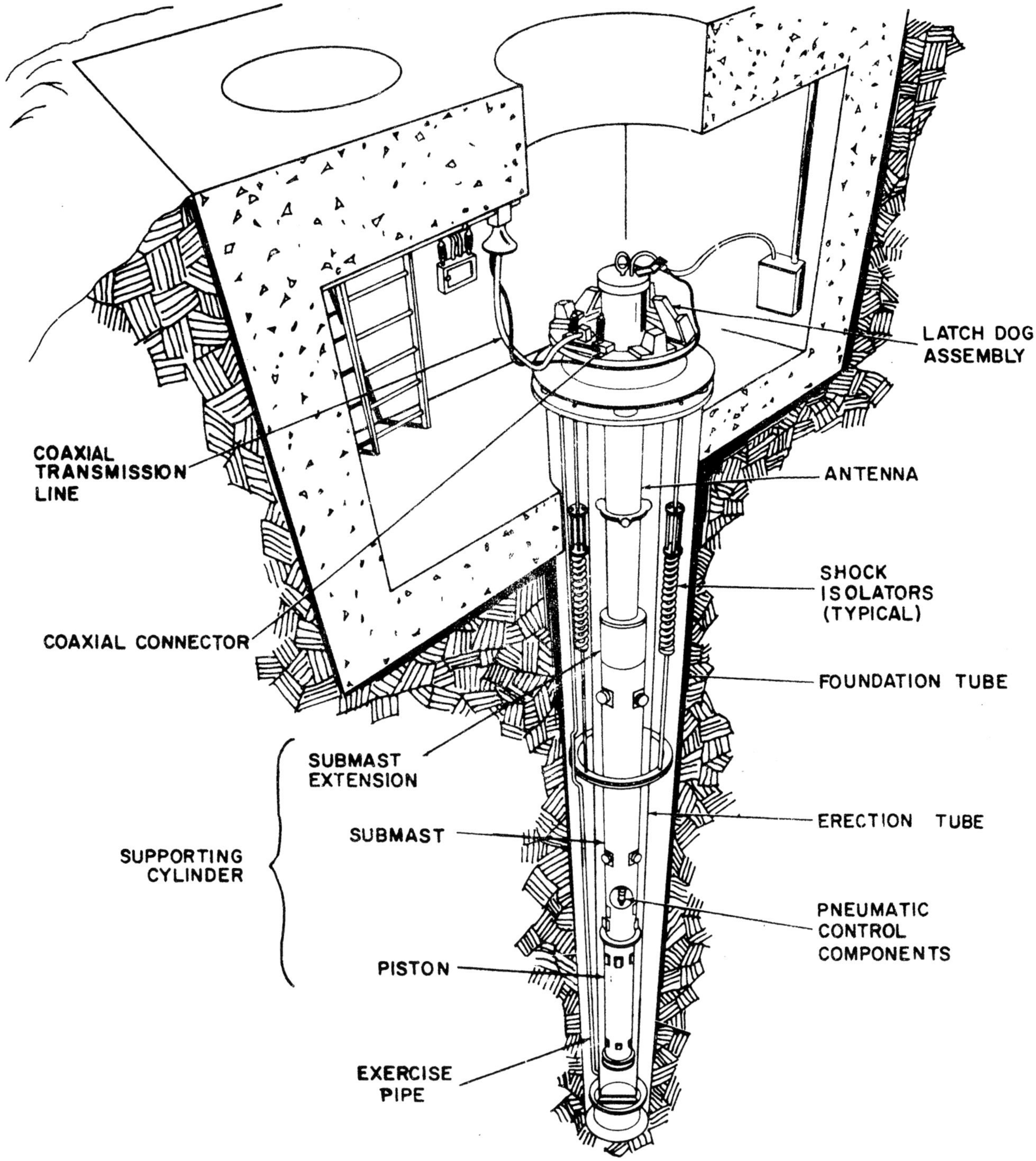

◀ Diagram of the hardened IRCS pop-up antenna and its lifting mechanism.

◀ UHF antenna, made by Collins Radio, with author for scale.

a button located in the LCC. This change was made to protect the antennas during the suction phase of a nuclear detonation. After the passage of a shockwave from a nuclear detonation, air is sucked back toward the low-pressure zone created at the center of the explosion. This causes high velocity winds traveling in the opposite direction of the original shockwave to sweep back over the area previously struck. Raising the backup antennas manually protected them from premature deployment and damage by the backwash. Once raised, there was no provision for lowering the antennas—had they been needed because of a

▶ The radio-type maintenance network (RTMN) topside antenna, with a voice signaling system (VSS) speaker.

▲ A portion of the RTMN slotted coaxial antenna, made by Andrews Microwave.

▲ The survivable low frequency communication system (SLFCS) antenna, as displayed at manufacturer's plant.

DECO ELECTRONICS PHOTO, TMM ARCHIVES.

nuclear detonation, there would have been no reason to lower them. When they were raised for testing purposes, the antennas had to be lowered by a maintenance crew. Unlike the HF antenna, the pop-up antennas did not telescope. Hence, the silos for the pop-up antennas are nearly 70 feet (21 meters) deep. There was a small cylindrical vault just beneath the surface to accommodate maintenance crews. Identical pop-up antennas are also found at the Mount Lemmon mountaintop relay site.[11]

UHF (Semi-Hard) Antenna

The semi-hardened cone-shaped ultra high frequency (UHF) antenna provided communications with helicopters and other aircraft associated with the wing command post. At launch sites around Kansas and Little Rock, this antenna could be used to listen for an airborne command post known as Looking Glass. But Looking Glass never flew far enough south to communicate with sites around Tucson.[12] Tucson launch sites used the same antenna to listen for the Emergency Rocket Communications System (ERCS). ERCS used a pair of Minuteman missiles based at Whiteman AFB in Missouri. If needed, these missiles could be launched by Looking Glass and would broadcast the launch order from transmitters in their nose cones.[13] Technical manuals describe the antenna as a "monopole radiator" in the shape of a "solid cylindrical probe weighing 200 pounds (90 kg), insulated from a heavy antenna base weighing 1700 pounds (700 kg)."[14] The "semi hard" description refers to the antenna's less-than-indestructible construction: while the solid metal antenna was intrinsically sturdy, the cone-shaped enclosure was simple fiberglass and would have been unable to protect the antenna from falling objects, flying debris, extreme heat or other nuclear effects. The antenna's concrete platform was 10 feet (3 m) in diameter and four feet (1.2 m) thick. Rough calculations suggest it weighed about 40,000 pounds (18,000 kg).[15]

Survivable Low Frequency Communications System Antenna

This antenna was installed in 1967 and was buried underground near the LCC. The antenna was made of two hoops, each about six feet (2 m) in diameter, and positioned perpendicular to each other with a north/south loop and an east/west loop. While there was nothing "hard" about the construction of the antenna, its underground location made it virtually impervious to the effects of a nuclear detonation.

The system provided receive-only data (not voice) communications for the survivable low

◀ Theodolite station. Note pads for locating tripod.

▶ Silo air exhaust shaft.

◀ At 571-7 the control center air intake shaft is now topped with an evaporative cooler. Note tipsie surveillance units.

frequency communications system (SLFCS, usually pronounced as "slifix").

Radio-Type Maintenance Network Antenna

Last but not least was the radio-type maintenance network (RTMN) antenna. The RTMN antenna was located on top of a pole next to the access portal. This soft antenna enabled communications between the control center and crewmembers or maintenance teams working at various locations topside. The antenna could be disabled when not in use. A special RTMN cable antenna was woven through the underground portions of the complex to ensure clear communications underground. Such an antenna was required because radio signals could not penetrate the heavy steel and concrete walls of the complex. Without the cable antenna, those in the LCC would have been unable to communicate by radio with those anywhere else in the underground portion of the complex. The cable antenna originated on level 3 of the control center and wound its way to the bottom of the silo. The RTMN operated near 140 MHz.

For more information on all communications systems refer to Chapter 5 and launch control center levels two and three.

OTHER TOPSIDE EQUIPMENT

Theodolite Station

The theodolite station was located at the northwest corner of the silo. A hatch could be opened to reveal a pipe slanting down to silo equipment level two. It was used in conjunction with a theodolite and two reference monuments

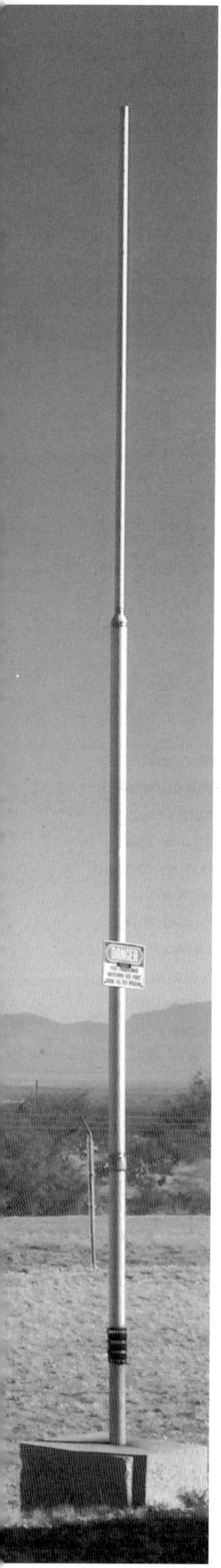

(located several hundred yards (meters) outside the security fence) to calibrate the guidance system's azimuth alignment set (AAS) autocollimator. For more information on the AAS refer to the section describing silo level 2.

Air Shafts

Three shafts provided air circulation for the underground complex. Two of these shafts accommodated fresh air intake and one was for air exhaust. The shafts were constructed of corrugated metal pipe sections three feet (1 m) in diameter and opened to the surface under a grating surrounded by a bright yellow railing.

One of the intake shafts was located near the LCC and supplied fresh air for the crew. The shaft also doubled as an emergency exit and was protected from intruders by a pair of tipsie units. At 571-7 the museum has installed an evaporative cooler at the top of this shaft, as the original system of underground fans and air conditioners is no longer operational.

The other intake shaft provided fresh air for the silo and launch duct and was located at the southwest corner of the silo.

The silo air exhaust shaft was located at the northwest corner of the silo and was marked DANGER TOXIC VAPORS because of the possibility of fuel or oxidizer vapors or diesel exhaust being expelled from the silo. Neither of the silo air shafts could be used for escape. Ladders in these shafts permitted only servicing of sump pumps at the bottom.

Another shaft was located at the northeast corner of the silo. This shaft provided access to a rupture disk designed to shut off the flow of water from the above-ground water reservoir in the event of shockwaves traveling through

◀ Rupture disk access shaft. A short ladder led down to a small vault providing access to the rupture disk itself.

◀◀ This vent stack for fuel vapors was used only during PTS operations.

▼ A specialized fuel incinerator was developed specifically for deactivation of the launch sites.

USAF PHOTO. TMM ARCHIVES.

▲ Portable oxidizer burn unit. Oxidizer vapors came in the bottom through fixed piping and were mixed with propane through a fitting on the right side. The black box contained a spark igniter transformer.

▲ The burner in operation.

the water pipe. The rupture disk was eventually abandoned in favor of a plug valve (discussed later).

Fuel and Oxidizer Stacks

The propellant transfer system (PTS) employed a closed loop of tanks and pipes to minimize the venting of fuel and oxidizer vapors while loading or unloading propellant on the missile. Still, after PTS operations were completed, it was necessary to purge the pipes involved in the process. The fuel vent stack was located northeast of the oxidizer hardstand. It was just a vertical pipe some 20 feet (6 m) tall. Vapors resulting from fuel line purging were routed to this tall pipe and vented to the atmosphere unburned. For deactivation of the sites, a special incinerator unit burned the vapors in a contained vessel. Exhaust from this unit was routed to the fuel vent stack through fittings spliced into the base.

Vapors from the oxidizer lines were mixed with propane and burned from a flare stack just southwest of the oxidizer hardstand. The burner unit was brought to the complex only as required.

Status Lights and Klaxon

Red, yellow and green lights on a pole near the silo door provided a visual indication of the status of the complex. The yellow and green lights were standard incandescent lamps but the red light was a rotating beacon. Under normal circumstances the green light would be illuminated indicating there were no special conditions or hazards on the complex. Under more unusual circumstances, during a PTS operation for example, the yellow light was illuminated to indicate potentially hazardous conditions and warn that nonessential personnel should stay away. The red beacon could, if desired, be accompanied by a siren mounted on the same pole, and was used to indicate emergency conditions or that a launch was imminent.[16]

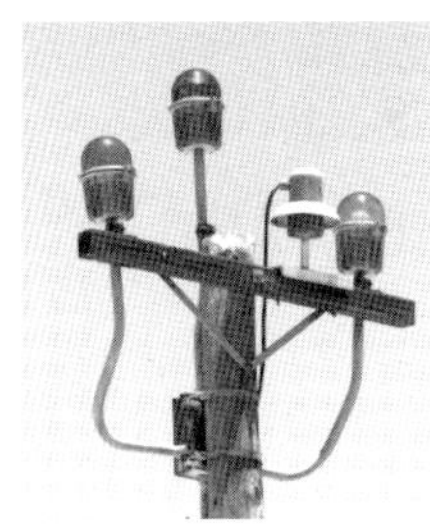

Thunderbolt Siren

A large air-powered siren was installed in 1981 in response to emergency events at a Titan II complex near Little Rock, Arkansas. Some readers will be familiar with these sirens, as they are still used in many parts of the country to warn of severe weather. During the Cold War thousands of cities and towns were outfitted with similar sirens to warn of impending nuclear attack or an air raid (an attack by bomber aircraft). Many of these sirens were made by the Federal Signal Corporation. The bright

▶ The Federal Thunderbolt. This air-powered siren could wake the dead.

◀ The dump tanks were abandoned and their pipes cut off and capped.

yellow "Thunderbolt" model was extremely loud, bordering on painful when heard from the complex, and could be heard very clearly more than a mile away. The siren was activated from a control box located in the LCC, and was tested every Friday at noon.[17]

The museum's Thunderbolt is still operational and is occasionally run for special events. It is rare treat and a hair-raising experience.

▶ Filler pipes for the diesel fuel tank. The pump transferred fuel to a tank in the silo. That no attempt was made to protect the transfer pump is a little curious. It was unlikely to survive an attack.

Diesel Fuel Tank

An 8000-gallon (30,200 l) buried tank stored diesel fuel for the emergency generator. The tank was protected to some extent by virtue of being underground, but was not specifically hardened against attack. Survival of the tank and its transfer pump was unlikely in any relatively close strike. Fuel from this tank was pumped as required to maintain 2500 gallons (9500 l) of fuel in a tank located in the silo.

Dump Tanks

Until 1979 there was a provision for pumping contaminated water from the silo sumps (in case of an in-silo propellant spill) or runoff from the hardstands (again as the result of a spill) into two large underground dump tanks. The dump tanks would also have been used in an emergency unloading of the missile's fuel and oxidizer. Once safely in the dump tanks the propellant was diluted with water to reduce the fire hazard and pumped into trucks for disposal. The tanks operated in conjunction with a total of seven valves (three in the valve pit and two at each hardstand). These valves directed flow from the silo sump pumps or hardstand drains to one tank or the other, or to surface drains. Emergency unloading of the missile's propellant involved a separate dedicated system of pipes.[18] In addition to the tanks having a history of leaking, there were a number of reasons they might not have worked very well. These include the highly volatile and corrosive nature of the propellants (even when heavily diluted with water), difficulty in purging the pipes and tanks, the potential to accidentally mix fuel and oxidizer by incorrectly setting the valves, and the need to connect hoses to the missile to unload the propellant. Exactly how this last feat would have been accomplished is unknown, but there were formidable obstacles. It is not known if the tanks were ever actually used. Concrete platforms mark the location of

▼ Two of the three valves in this pit were removed after the dump tanks were abandoned. Valves routed contaminated water to one of two tanks or to surface drains. Turn these the wrong way and you could dump fuel into the oxidizer tank or visa versa.

the tanks. The fuel dump tank, typically 60,000 gallons (225,000 l), was located a few yards (meters) southeast of the fuel hardstand, while the oxidizer tank, typically 20,000 gallons (76,000 l), was located a few yards (meters) to the west of the oxidizer hardstand (see diagram on page 13). The platforms once had pipes for venting and draining the tanks, and large dial-type gauges to display the level of fluid within. They were abandoned in place in 1979.[19]

Management of propellant spills was a complex matter. Three factors played into the decision on how to deal with a spill: 1) The size of the spill, 2) the location of the spill, and 3) the type of spill (fuel or oxidizer). Dilution with water was almost always required, but diluting a large quantity of oxidizer was tricky because mixing it with water generated a tremendous BRFC, jepordizing downwind civilians.[20] In some cases dilution with water at a ratio of 28:1 could be required. For a sizable spill this would result in a huge volume of liquid, making both containment and disposal a serious problem.

Transformer and Cooling Tower Pits

Between the Thunderbolt siren and the silo closure door were two pits: one enclosed with a metal cover and the other open and surrounded with a handrail.

The covered pit contained a massive electric power transformer. This transformer was connected to the local electric utility lines and provided the complex with 480-volt, 3-phase electricity (think of "3-phase" as industrial-strength electricity). The transformer was connected directly to a large distribution center on silo level 3. If commercial power failed or went out of tolerance an emergency generator, also located on silo level 3 could take over. Additionally, a junction box in the transformer pit permitted connection to a mobile generator which could be delivered to the complex if needed.

Adjacent to the transformer pit was the cooling tower pit. A modification made only in Tucson replaced the original cooling towers with more modern refrigerated coolers in an effort to reduce water consumption. The original cooling towers worked in conjunction with the site's underground water chillers to expel heat from the complex, helping to maintain the launch duct and missile at 60° F (15° C) plus or minus two degrees F, and the rest of the complex at around 72° F (22° C). The modern coolers did not require the underground chillers except during plumbing shutoff conditions (discussed later) during an advanced stage of readiness.

◀ This large transformer converted commercial power to 480 volts.

◀ The now-empty cooling tower pit. Note covered transformer pit with ventilator in background.

◀ Cooling towers as seen during construction. The units used copious quantities of water to expel waste heat from the complex.

▶ In Tucson, modern chillers made by Carrier were installed in 1980. These greatly reduced the amount of water used by the complex.

USAF PHOTOS, TMM ARCHIVES

Delta-T Pole and Weather Vane

A 55-foot (17 m) wooden pole topped with a platform was used in conjunction with a nearby wind speed and direction indicator, casually referred to as the "Red Baron" (a reference to the famous WWI pilot because of the unit's red color and resemblance to an airplane). The delta-T and Red Baron were used to assess weather conditions prior to and during propellant transfer operations. A PTS operation was permitted only when the wind was from a particular direction, and when wind velocity was between an established minimum and maximum speed. Further, a propellant transfer was not permitted when inversion conditions existed. An inversion is a reversal of normal temperature gradients above the surface of the earth. Inversions are responsible for trapping exhaust fumes from cars, causing the brown layers of smog all too familiar to motorists. In the same way, inversions would prevent the dispersion of toxic vapors that might be present during a propellant transfer.

USAF PHOTO. TMM ARCHIVES.

▲ One of two thermometers used to check for inversions.

A pole with the platform on top, called the delta-T pole, was equipped with a thermometer at the top of the pole and another a few feet off the ground. The temperature difference between these two thermometers (called the "delta-T") was recorded on a chart in the LCC and indicated if an inversion was present.

▼ A summer monsoon storm gathers beyond the Red Baron and Delta-T weather equipment.

Wire-Type Maintenance Network

The wire-type maintenance network (WTMN) served much the same purpose as the RTMN: it was used to communicate with crew members and maintenance teams working both topside and in the underground portions of the complex. Dark blue pole-mounted boxes at several topside locations contained receptacles into which headsets with microphones could be plugged. A "call" button inside the boxes enabled the user to signal the LCC. The WTMN operated as a "party line," meaning that all users could hear and talk to all other

◀ Typical wire-type maintenance net (WTMN) connection box. Note silo door rails.

▶ Inside the WTMN boxes were "call" buttons and connectors for headsets.

users simultaneously.[21] Underground, WTMN headphone connector jacks were located throughout the silo equipment area and launch duct as well as in the LCC.

Voice Signaling System

The voice signaling system (VSS) was what the rest of the world would call a public address system. Announcements over the VSS were made from the LCC by pressing the VSS pushbutton on the communications panel and speaking into the telephone handset. VSS speakers were located throughout the complex both underground and topside. As one might expect, the VSS was used to make announcements to personnel on the complex, warning of hazards, for example, or what action should be taken.[22] Such announcements invariably began with the words "Attention on the complex! Attention on the complex!"

Breakaway Fence Sections

Exiting the topside area of the complex normally required contacting the LCC from the phone at the main gate. The LCC would then unlock the gate, allowing it to be rolled open by hand. In an emergency, however, when time was of the essence, being able to get away from the complex in a hurry could be critical. To facilitate a hasty departure, several sections of the perimeter fence were fitted with special poles and quick release pins that, when pulled, enabled those sections of fence to open. These poles were painted red for easy identification.

True North Pole

A red and black striped pole just west of the silo is identified on some blueprints as the "true north pole." While no specific reference describing it function has been found, it is believed to have been used to calibrate the site's wind direction indicator, as the pole is located exactly north of the wind speed and direction equipment.

▶ The "true north pole." Santa has never been sighted.

▲ Typical WTMN headset.

Manholes and Handholes

Manholes and handholes differed from each other principally in size. Manholes were large enough to contain equipment and a number of people. Handholes were quite small and contained only enough room for a junction box and an individual. There were several handholes on the complex, but only two manholes.

Manhole J was a blast-hardened junction box feeding underground cables to the IRCS pop-up and HF hardened antennas. It contained EMP and lightning protection equipment and typically was located very near the pop-up antennas. Manhole J appeared as a rectangular concrete pad with an access cover near one

▶ Manhole J. A hardened vault provided connections to the hardened antenna systems. The seven small white plug marks are seen at all site ruins. They suggest some modification, but no reference to it has been discovered.

▶ Entry to the PTS receptacle pit was via the two hatches under the PTS control trailer. The foreground handhole provided access to VSS and WTMN connections.

▶ Typical handhole. There were several around the complex. These contained junction boxes for a number for soft systems.

corner. The cover opened into a vault about 5 x 7 feet (1.5 x 2 m) in size. The designation "manhole J" is curious since there were no manholes A though I.

The only other manhole was located between the fuel and oxidizer hardstands (near the engine exhibit at 571-7) and opened into the PTS receptacle pit. The 5 x 7 foot (1.5 x 2 m) vault had two red access covers and contained a junction box for connecting communication and control cables from the PTS control trailer. The PTS control trailer was a kind of mobile command center that coordinated all activities during propellant transfers. The PTS receptacle pit was not hardened.

Handholes provided access to soft radio antenna cables, VSS and WTMN lines, electrical junction boxes and a variety of other expendable systems. They were fitted with familiar looking street-type manhole covers. The covers were painted red and had the word "communications" molded into the top, and opened to reveal a 4 x 4 foot (1.2 x 1.2 m) vault. Handholes were not hardened and were located at several places around the complex.

Water Supply and Soft Water Storage Reservoir

The sites consumed prodigious quantities of water— each one used an average of 2500 gallons (9500 l) daily. Water for most launch sites, including 571-7, was supplied from deep wells, and a few sites had more than one. The remaining sites were connected to commercial water mains because of their proximity to those sources. The location of the wells varied from site to site. At 571-7 the well was just outside the west fence near the IRCS soft antenna. Each complex had a topside water storage facility consisting of a large concrete-lined pit with sloping walls, called the soft water storage reservoir. This did not mean soft as in "water softener"; it meant the reservoir was not hardened to withstand a nuclear attack. Water from the well or commercial mains was treated with a chlorinator and softened (if necessary), then pumped into the

◀ This concrete-lined reservoir held 100,000 gallons of water. It was originally covered only with a wood and tarpaper roof.

soft water reservoir. The reservoir held 100,000 gallons (380,000 l) of water, which was pumped as required into an underground tank located in the silo. Water from the reservoir also served the site's domestic water system for drinking and cooking. With that in mind, it should be noted that water is a magnet for wildlife in the desert southwest, and it is reasonable to assume that since the reservoirs were covered only with wooden planks and tarpaper, all manner of desert creatures found their way into the reservoirs, many drowning in the process. One hopes that copious amounts of chlorine were used. A filter would have been a good idea as well. But while "strainers" are mentioned in technical documents there is no mention of a filter, as that word is commonly understood. Many crew members refused to drink the water, preferring to bring their own.

◀ The well at 571-7. The museum is now connected to commercial water mains.

Sites in Kansas and Arkansas used a reservoir of a different design and probably less subject to infiltration by wildlife. As of this writing, stories of crew members swimming in the reservoirs remain uncorroborated.

1 Sylvania Electronic Systems, Information Bulletin, *Atlas-Titan Radar Surveillance Security System AN/TPS-39*, 1964. TMM archives

2 Col. Hugh Matheson USAF (ret.), Lt. Col. Orville Doughty USAF (ret.), interview

3 1Lt. Ed Moran (ret.), video of his tour for museum docents, circa 1998, TMM archives

4 Brad White, former Air Force communications technician, interview

5 Sylvania Electronic Systems, Information Bulletin, *Atlas-Titan Radar Surveillance Security System AN/TPS-39*, 1964, 9. TMM archives

6 U.S. Army Corp of Engineers, *Titan II Dismantlement Blueprints*, Little Rock AFB, Sheet C-28

7 TO.21M-LGM25C-1, 1-116G

8 ACPs were launch sites equipped with additional communications equipment and could assume command if the WCP was destroyed.

9 390th DOTI, *IRCS*, 1977, TMM archives

10 TO.21M-LGM25C-1, 1-116G

11 Author's research at Mt. Lemmon site

12 Maj. George Birch USAF (ret.), interview

13 Daniel Ford, *The Button*, (Simon and Schuster, 1985), 150

14 TO.21M-LGM25C-1, 1-117

15 Author's field research

16 Ogden Air Logistics Center Pamphlet 00-ALCP 144-4, Fuels, Propellant and Chemicals, *Titan II Propellant Hazard Management Guide*, 1983, A3-35

17 Letter, dated October 1981 to area residents regarding installation and testing of sirens, from 390th SMW commander Col. Paul Comeaux. TMM archives

18 *Weapon System Familiarization*, Course OZR1821F/3121F, Technical Training, Shepard AFB, January 1965, 157, 162

19 Facility Engineering Change MCL 3263, TMM archives

20 00-ALCP 144-4, 4.50-4.61

21 TO.21M-LGM25C-1, 1-87, 1-88

22 TO.21M-LGM25C-1, 1-99

▼ The access portal was the main entrance to the underground complex. Unlike the rest of the complex, the portal was not expected to survive the effects of a nearby nuclear hit. When not in use the elevator was always kept below ground.

CHAPTER 4

GOING UNDERGROUND

Although it was built with steel reinforced concrete walls more than a foot (30 cm) thick, the access portal was regarded as soft. If a nuclear bomb exploded nearby it is likely the structure would have collapsed.

The access portal was approximately 16 feet (5 m) square and about 35 feet (10 m) deep. Major components of the portal included the outer stairwell and stairwell cover (usually left open); the outer portal door and phone; the entrapment area, telephone and closed-circuit TV camera; the stairway; the elevator; and a blast door with its associated phone. Also located within the access portal was an exhaust fan with its vent hood on the "roof" of the portal and a sump pump located at the bottom of the elevator shaft.

Signs both inside and outside the outer portal door at sites near Tucson warned the unsuspecting to watch for rattlesnakes. Rattlesnakes were a danger unique to the Tucson sites and are still a hazard today. Now and then, especially during the hot summer months, snakes would slither down the steps to keep cool on the shady concrete at the foot of the access portal door. This could come as a big surprise, especially for those exiting the complex. Snake-handling sticks were standard equipment.

Personnel arriving at the complex would first call the LCC from the perimeter gate phone. The gate was then "buzzed" open from the LCC and rolled clear by an oncoming crew or maintenance-team member. A second call was then made from a phone just outside the access portal outer door. Another security detail: there was a three minute time limit between the first and second phone call! When the access portal door was buzzed open, personnel entered

► The underground portion of the launch complex was composed of four separate structures: the access portal, the blast lock, the control center, and the silo. Two steel tunnels, called cableways, connected the silo to the control center by way of the blast lock, while the blast lock was connected to the outside world via the access portal.

USAF PHOTO, TMM ARCHIVES.

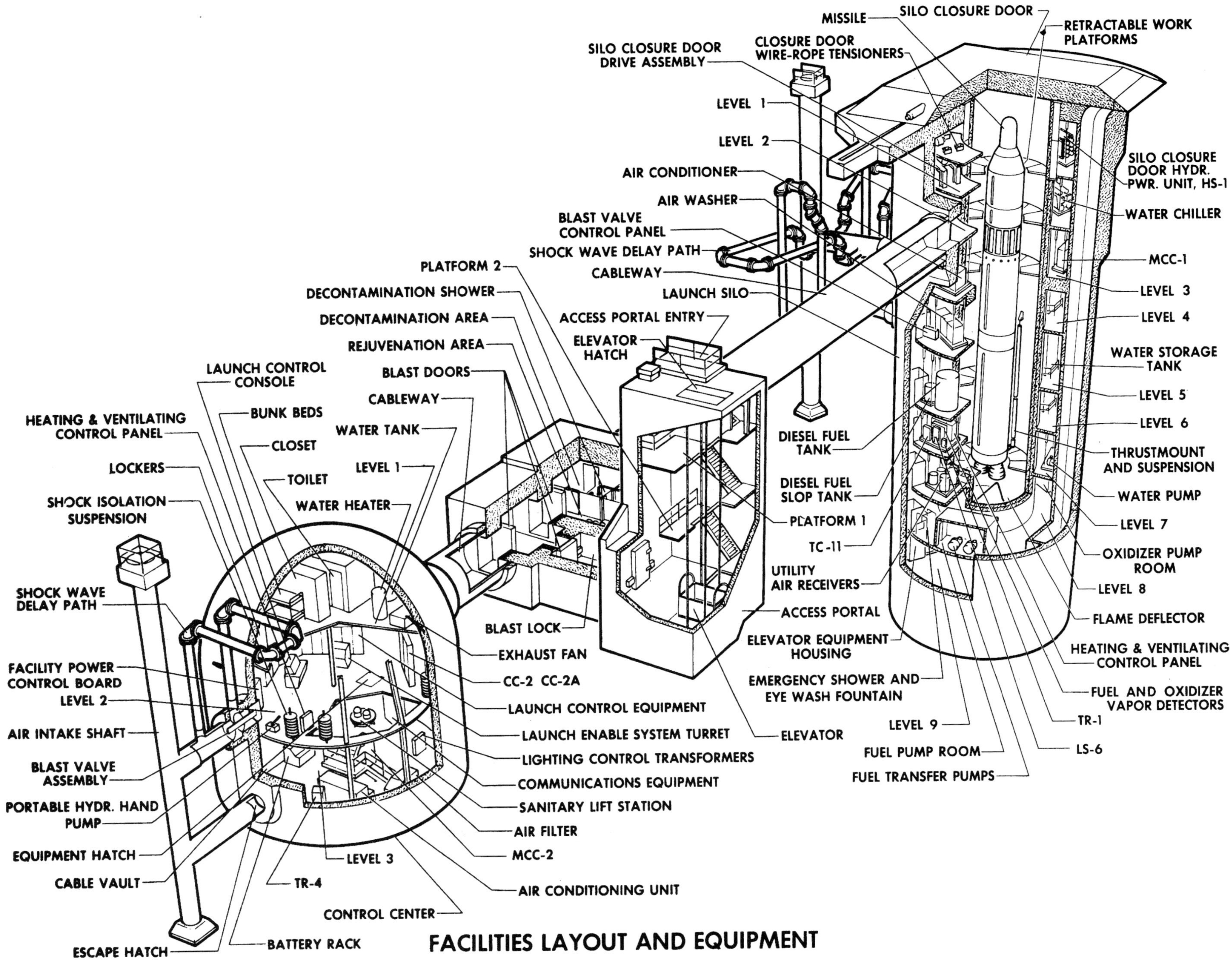

FACILITIES LAYOUT AND EQUIPMENT

TMM ARCHIVES.

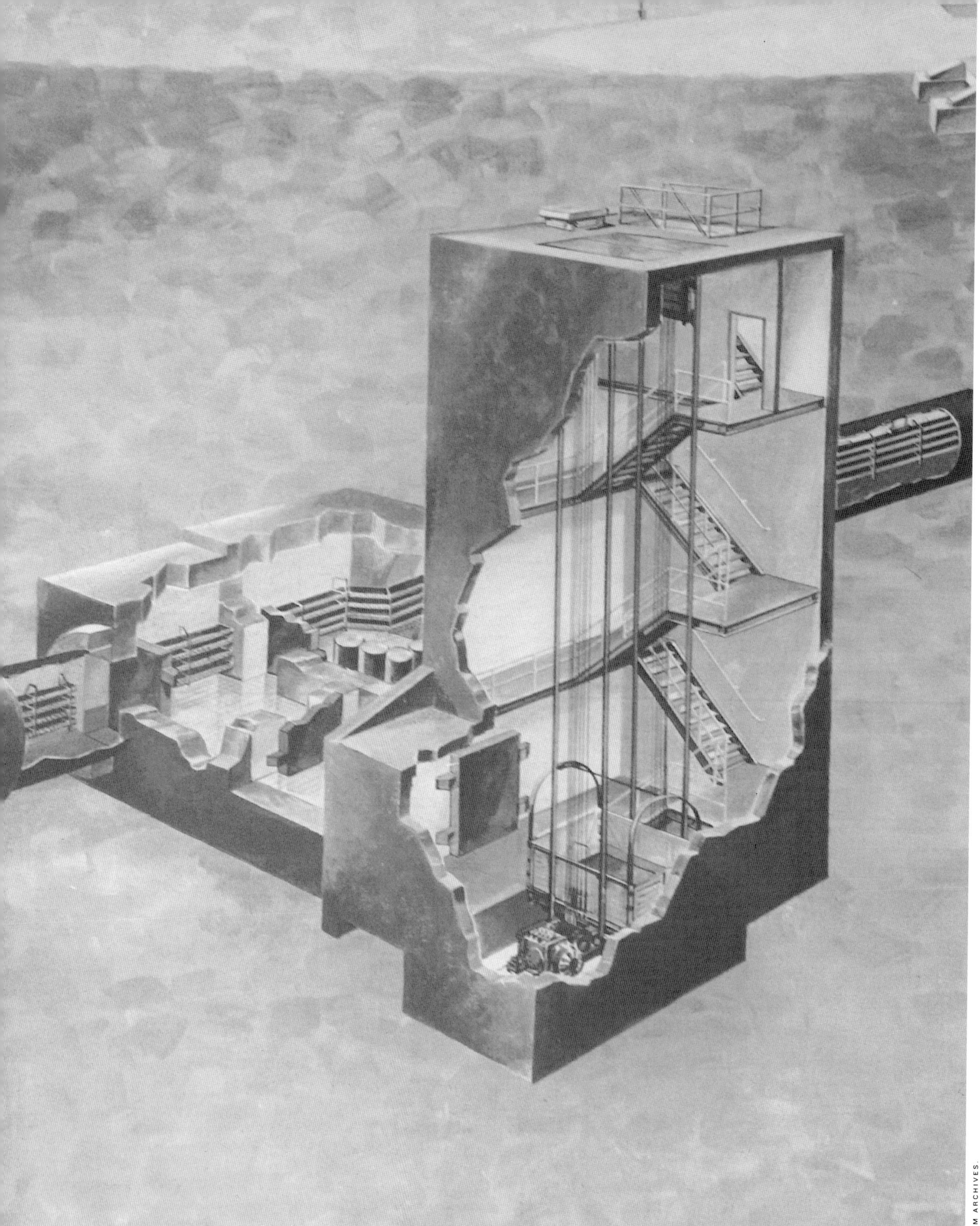

TMM ARCHIVES.

▲ Looking into the entrapment area from the outer portal door. Note TV camera.

the entrapment area, which confined those entering until their authorization to enter could be confirmed. The entrapment area consisted of two electrically locked doors, one at each end of a short stairway. When the outer door closed, those inside were trapped between it and the inner door. Both doors had to be buzzed open with a button in the LCC, but both doors could not be unlocked at the same time.

To get through the second door another phone call was made to the LCC and a six-letter code recited. Each code was used only once—everyone (or every group) coming through the entrapment area had a new code, and the code card was burned after use. There is an oft-told legend about having to eat the code card if one did not have matches or a lighter. It is conceivable that this may have happened once or twice as part of a prank, but eating the code was not an approved method of disposal and all

▼ Access portal stairway leading to blast door 6.

▲ Looking up from inner door. Note telephone and VSS speaker at top of stairway, and burn can at right.

FOUO

Ø	1	2	3	4	5	6	7	8	9
H	I	J	K	L	M	N	O	P	Q
Ø	1	2	3	4	5	6	7	8	9
O	P	Q	R	S	T	U	V	W	X
Ø	1	2	3	4	5	6	7	8	9
D	E	F	G	H	I	J	K	L	M
Ø	1	2	3	4	5	6	7	8	9
P	Q	R	S	T	U	V	W	X	Y
Ø	1	2	3	4	5	6	7	8	9
Q	R	S	T	U	V	W	X	Y	Z
Ø	1	2	3	4	5	6	7	8	9
W	X	Y	Z	A	B	C	D	E	F

USKTP 78 ØØØØ3
Ø128

TMM ARCHIVES.

▲ Small code cards, issued by Code Control (located on the Base), were essential in gaining access to the underground portion of the complex. "FOUO" means "For official use only." After use the cards were burned in a red-painted coffee can. Forget the stories about having to eat the card if you didn't have any matches, but getting the code wrong meant you were going to have a bad day.

▲ It took about 70 seconds to make the 35-foot trip up or down. Note crane assembly for heavy objects. The ramp and handrails were added by the museum.

procedures involving the code were taken very seriously.[1] As an additional security measure, those in the entrapment area were observed on a closed-circuit TV camera. Once past the entrapment area, one walked down the stairs and made the last call from the phone at the blast door.

The access portal elevator was used to haul heavy or bulky items and other supplies that would be difficult or impossible to manage on the stairway. The elevator could not be summoned from the surface. One had to clear the entrapment area and bring the elevator up from below, but only after power to the elevator had been turned on by the crew in the LCC.

At the top of the access portal, the elevator shaft was covered by hatches that served primarily to keep out the weather. The hatches were pushed open by a curved frame on the top of the car, and as the elevator approached the surface an alarm bell sounded to warn those topside that the hatches were about to open.

Blast Lock

The blast lock was the interface between the outside world and the underground launch complex. It was located adjacent to the access portal and between the LCC and silo structures. Pairs of massive steel blast doors and immense concrete and steel walls protected the crew from intruders, the effects of nuclear weapons—and their own missile.

The blast lock was constructed by pouring three-foot (1 m) thick concrete walls inside a quarter-inch (6 mm) thick steel shell. Walls with blast doors were four feet (1.2 m) thick to compensate for their more direct exposure to potential overpressures. The walls were reinforced with steel rebar as large as 1½ inches (4 cm) in diameter. The floors and ceiling of the blast lock were about five feet (1.5 m) thick and similarly reinforced.[2]

The blast doors were made of 12-inch (30 cm) I-beams. Each door was 5½ feet (1.7 m) wide and 7½ feet (2.2 m) tall. In 1995, David Stumpf used blueprint drawings to calculate the weight of the doors and arrived at a figure of about 6000 pounds (2700 kg).[3] Titan Missile Museum docents universally use that number. Some faint shadow of doubt was cast on that figure in

◄ Blast door 6, the "front door" of the complex. Note locking pins on the door frame, and sockets on door edge. Locking and unlocking pushbuttons are at left of door frame.

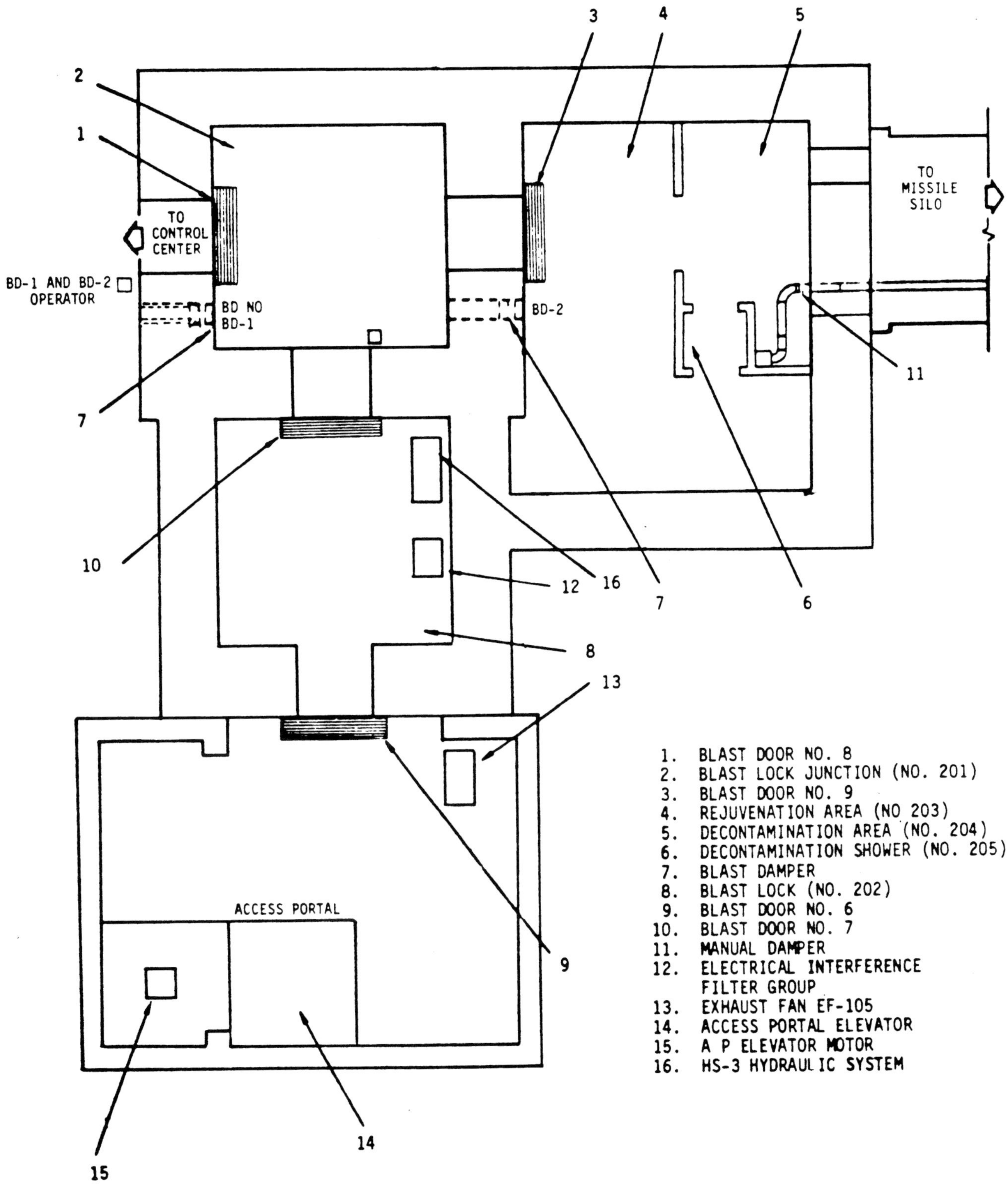

TYPICAL ACCESS PORTAL AND BLAST LOCK

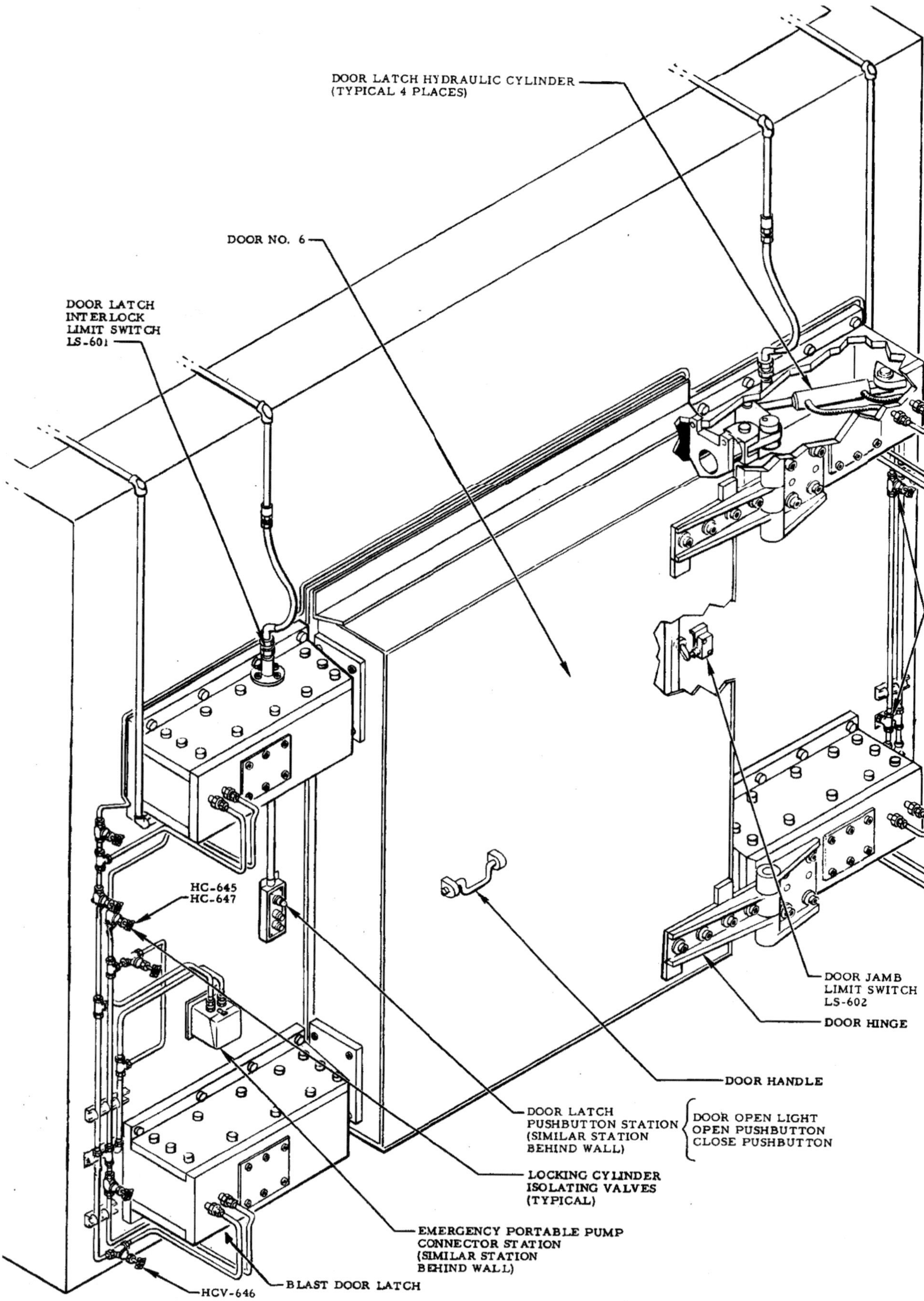

◄ Blast door locking pins were operated by an elaborate hydraulic system. Note limit (interlock) switches and connection box for emergency hand pump.

TMM ARCHIVES

▲ A view from door 7 to door 6 and the access portal beyond. Note latch on wall at right of door and eye-hook on the door itself.

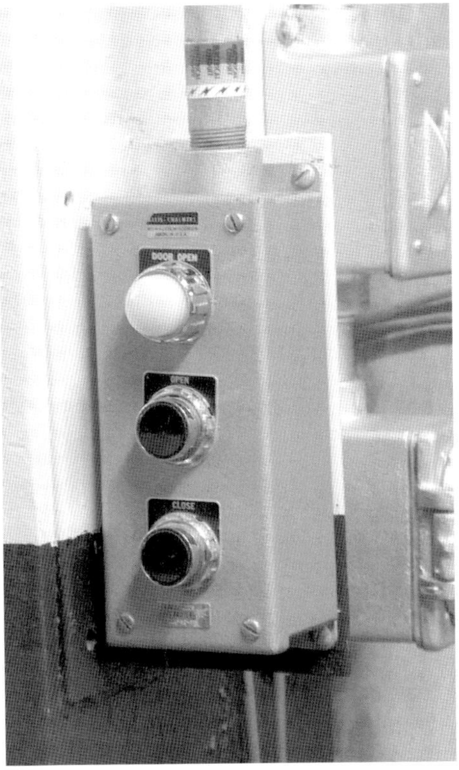

◀ Typical blast door open/close switch. The light indicated when the door was unlocked.

2007 when the museum's archivist discovered a 35mm photographic slide showing one of the blast doors with a stenciled sign reading "DOOR WEIGHT 7000 LBS. DO NOT SLAM." It is not known if the Air Force's figure was any more reliable than Stumpf's, but regardless of the actual weight, the blast doors were both impressive and formidable.

The doors were not counter-balanced or spring-loaded, yet could be moved with remarkable ease on perfectly balanced hinges. The doors and doorjambs were pre-assembled and emplaced as a single unit (see photo on page 26). The total weight of the door/jamb assembly was about 37,000 pounds (16,800 kg). The doors and jambs had machined surfaces, and the jambs were fitted with seals so that when the doors were closed, the assembly was airtight.

In designing the doors for the blast lock, engineers considered two scenarios: a nuclear blast wave traveling down the access portal and a fuel-oxidizer explosion of the missile in the silo. While the rest of the complex, including the silo door, was designed to handle overpressures up to 300 psi (21 kg/cm^2), calculations suggested that blast waves traveling down the access portal could reflect and refract in ways that could actually magnify their force. So the blast doors were designed to withstand an overpressure of 1000 psi (70 kg/cm^2).[4] This level of protection also worked for the expected overpressure caused by an explosion of the missile in the silo. For perspective, most residential construction would be completely destroyed by an overpressure of just 5 psi (0.35 kg/cm^2).

The four blast doors were numbered 6, 7, 8, and 9. (There is no difinitive explanation as to why the doors were so numbered). Four large steel pins on the doors' frames fit into sockets on the doors' edges, locking the doors in place. Retracting the pins to open door 6 (the "front door" of the complex) required the cooperation of the crew in the LCC. Two buttons had to be pressed simultaneously: one at the door and another in the control center. The remaining three doors could be unlocked and opened simply by pressing the button at the door in question. An indicator in the LCC lighted whenever any of the four doors were unlocked. Whenever doors 6 or 9 opened, an audible alarm and warning light were activated.

If the hydraulic system used to operate the locking pins malfunctioned it was possible to unlock the doors with any of several strategically placed hand-operated hydraulic pumps. Because of this vulnerability, door 7 could be secured with what amounted to a giant "night latch"—a large hook and eye assembly that prevented the door from being opened under any circumstances. There is disagreement about the use of the latch. Some accounts suggest the latch would have been used only if there was an immediate threat to the complex, while others maintain it was to be latched every night.[5,6] Both accounts could be true since many procedures changed over the years.

▲ Original artwork on blast door 6 at 571-7, "the Copper Penny."

TMM ARCHIVES.

▲ Artwork at 570-4, "Liberty's Guardian."

TMM ARCHIVES.

▲ Artwork at 571-1, "Cochise Stronghold."

All in a Name

All launch sites were given numerical designations: 571-7 or 570-2, for example. But when crew and maintenance team members referred to a particular site they very often dropped the first two digits and the dash. 571-7 became "one seven" for example, while 570-2 became "zero two." Most former crew members still use this shorthand when talking about particular sites.

All 54 Titan II complexes had nicknames given to them by the crew (site 571-7 was the "Copper Penny" because of its proximity to a copper mine). Around 1980 crews got the idea (and the permission) to decorate blast door 6 with artwork reflecting the site's nickname. In some cases the artwork was quite elaborate. The door art was done by crew members, and a few sites had additional artwork on blast doors 7 and/or 8, and in rare cases, on the walls of the LCC.

TMM ARCHIVES.

▲ Artwork at 571-9, "The Observatory," named for its view of the observatory at Kitt Peak.

▲ HS-3 operated the blast door locking pins. It was located between blast doors 6 and 7.

The blast doors were equipped with electrical interlocks, and under normal circumstances both blast doors in a pair (6-7, or 8-9), could not be opened at the same time. It was possible, however, to override the interlocks, thus permitting doors 6, 7, and 9 to be opened together. This was done routinely for practical reasons: the transport of large, awkward or cumbersome equipment, for example, and during propellant loading or unloading operations to permit quick egress in the event of a fire or other internal calamity.[7] While it was possible to bypass the interlocks on door 8, which led to the LCC, regulations demanded this door remain closed at all times except to permit passage of crew members or maintenance teams.

Hydraulic System 3

Hydraulic system 3 (HS-3), one of four hydraulic systems on the complex, powered the blast door locking mechanism and also the blast dampers between blast doors 8 and 9 (discussed shortly). HS-3 was located inside the blast lock and at 571-7 is partially hidden behind blast door 7 (normally left open for visitor safety). Like the other hydraulic systems on the complex, HS- 3 incorporated the principle of pneudraulics, meaning that it used both pneumatic (pressurized gas) and hydraulic (pressurized fluid) systems to operate.

Shock Isolation

A nearby nuclear explosion would have sent powerful shockwaves through the earth, smashing into the complex. The massive blast lock walls could withstand shock waves, but it was important to protect critical equipment inside the walls from shock as well. In the blast lock and elsewhere, everything regarded as mission-critical was suspended from springs—a technique called shock isolation. Critical equipment included all the lights, electrical boxes, pipes and cables, and virtually all machinery. Additionally, shock isolated electrical devices were attached to flexible electric cables with plenty of slack to permit bouncing up and down without the cables tearing lose. Similarly, pipes were fitted with flexible sections allowing them to move and twist without breaking.

▼Typical shock isolation using spring canisters and flexible connections.

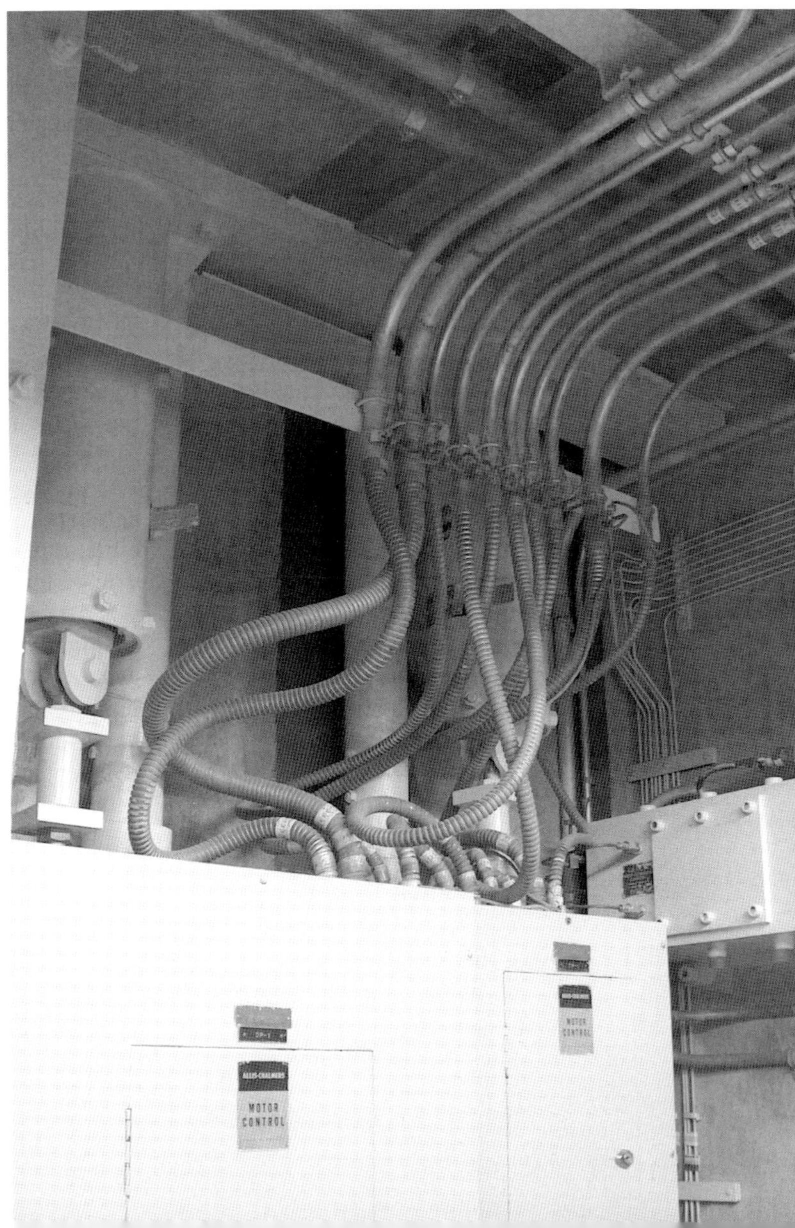

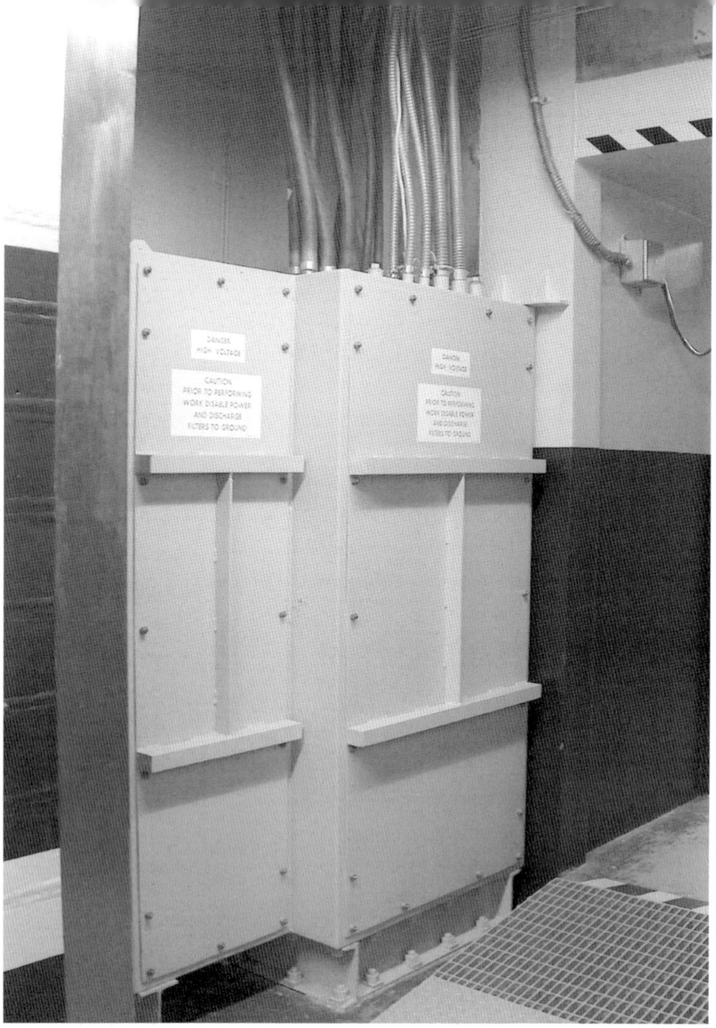

▲ EMP filters between blast doors 6 and 7.

Electromagnetic Pulse Protection

In addition to heavy concrete and shock isolation, another less obvious but equally important form of protection was in play on the complex.

Nuclear detonations, especially those high in or above the atmosphere, produce very intense pulses of radio energy called electromagnetic pulses (EMP). EMP can be thought of as a much more powerful form of the static caused by lightning and heard on a radio during thunderstorms. EMP is potent enough to burn out virtually any electronic device including radios, telephones, computers, electric wiring, motors, fuses and even light bulbs. EMP can cause damage in two ways. Behaving like ordinary radio waves, EMP can wash over sensitive electronic equipment, damaging the equipment directly. EMP can also pass through the wires of electric power lines, which act like antennas, inducing high voltage spikes that enter electrical equipment connected to the commercial electric power grid.[8,9]

Titan II launch sites were protected against both forms of EMP damage. A steel shell provided the first line of defense and protected the entire complex. At 571-7 part of this shell can be seen on the wall of the blast lock outside blast door 6, but is difficult to recognize because it has been painted over. When EMP waves strike this shell (called a Faraday shield), the currents they induce are absorbed directly into the earth surrounding the complex and cannot get inside. The complex is said to be grounded.

To protect against EMP voltage spikes entering equipment through the electric power lines, the complex employed the same kind of device you may use to protect your personal computer—the surge protector. These units are known officially as electrical interference filters, and more casually as EMP filters. They also employed lightning arrestors to protect the site and the missile from thunderstorms.[10] All electrical and radio cables entering or exiting the complex passed through EMP filters.

▲ Off-the-shelf emergency lighting units were located throughout the complex. These rechargeable units could provide light for several hours. Many were connected to extension lights that substantially reduced operating time.

Emergency and Warning Lights

Emergency lighting was provided by a number of stand-alone commercially made emergency lighting units (ELUs) suspended from shock isolation springs and located throughout the complex. These units had internal six-volt rechargeable batteries and turned on automatically in the event that commercial or generator power failed, and during the interval of time between loss of commercial power and backup generator start-up (generally less than two minutes).[11] Many ELUs were connected to extension lamps that provided illumination for areas not directly covered by an ELU box. These extension lamps were housed in shock-isolated explosion-proof enclosures with red protective cages over the lenses.

Engineers originally thought the backup generator would not simultaneously operate all the equipment needed to keep the missile ready to launch *and* the standard fluorescent lighting, so they designed the "diesel emergency lighting

▲ Auxiliary lights like this one (foreground) probably were never used. The warning light (background) turned on any time the warning klaxon sounded.

system" that activated automatically when the complex switched to generator power (a situation not uncommon in the early days of the program). When the diesel generator started, the site's main fluorescent lamps—together with a variety of normally autonomous "heavy load" equipment—were automatically turned off by permissive relays in order to avoid overloading the generator.[12] At the same time, other relays would activate a comparatively small number of standard incandescent lights. The diesel emergency lights were generally located near the extended lamps of the ELU system and were housed in explosion-proof enclosures with yellow protective cages. These would have provided a bare minimum of lighting on the complex. Key-operated override switches located in the LCC, blast lock, and silo equipment level 6 could be used to restore fluorescent lighting in those locations if needed.[13] As it turned out, the generator could easily handle the load of the fluorescent lights. As a result, the diesel emergency lights were officially abandoned in 1968.

Warning lights, similar in appearance to the ELU extensions and the alternate lighting system, could be found in several locations: on the access portal side of blast door 6, in the blast lock between door pairs 6-7 and 8-9, and near an equipment locker on the silo side of blast door 9. These lamps were enclosed in red lenses and lit up whenever the alarm klaxon sounded. Other warning lights were located near the diesel generator on silo equipment level 3 and outside the fuel and oxidizer pump rooms on silo equipment level 8.

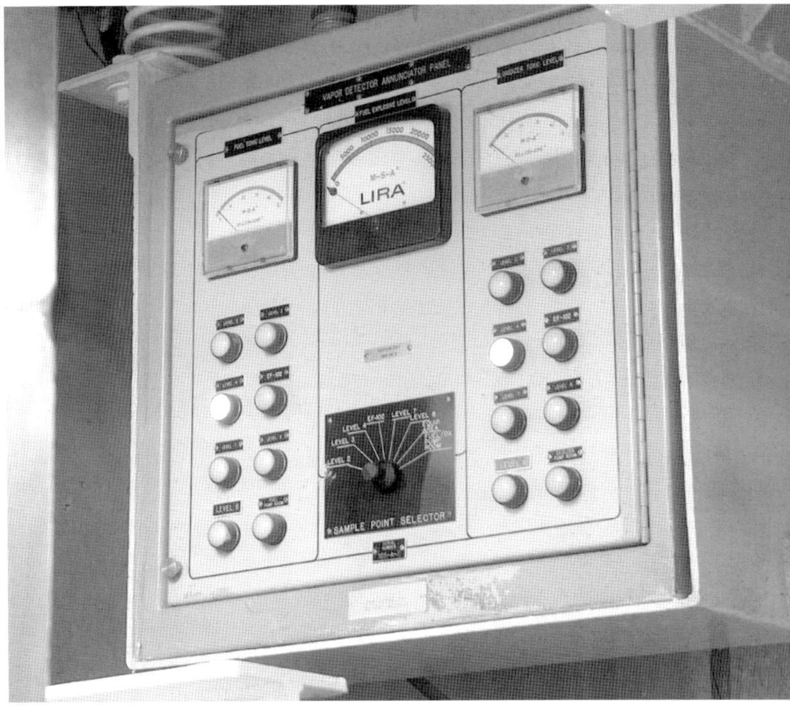

▲ The VDAP, located between blast doors 8 and 9, was used to check for propellant vapors in the launch duct.

Vapor Detection Annunciator Panel

Suspended from the ceiling near door 9 was the vapor detection annunciator panel (VDAP). The VDAP was the indicator panel for equipment that electronically "sniffed" for the presence of fuel and oxidizer vapors in the launch duct and silo equipment areas. The amount of such vapors registered on large meters on the VDAP, which was always checked prior to venturing beyond blast door 9. In 1983 a modification order was issued to relocate the VDAP to the control center.[14] At 571-7 the VDAP was not moved because the site had been deactivated before the order came through.

▲ Silo side of blast door 9. Blast damper (center) provided air flow from the control center to the silo via the blast lock. It would be pushed closed by a shock wave from the silo. Note emergency net phone (left), hand-pump connection box (below damper), and hand pump (lower left).

▲ Manual control for blast dampers, located on the control center side of blast door 8.

Blast Dampers

Between doors 8 and 9 were two blast dampers. The dampers allowed air to flow through the blast lock in only one direction: from the LCC to the silo equipment area. Airflow from the LCC was essential in providing fresh air for the cableways, and also in preventing toxic vapors and/or smoke from migrating out of the silo and toward the blast lock and LCC. In a nuclear attack or a chemical explosion of the missile, shock waves from the explosion would push the dampers closed, protecting the crew. The dampers could also be closed manually from controls located on the LCC side of blast door 8 (see diagram on page 51).

Emergency Net Phones

Five emergency net phones provided communications with the LCC for those trapped in the blast lock, or on the silo side of blast door 9. The phones were located between doors 6 and 7, doors 8 and 9, in the short cableway, on the silo side of blast door 9, and at the silo end of the long cableway. The phones had no dials. When picked up they automatically rang the phones in the LCC. The emergency net phones did not function as a party line. If more than one phone was in use, a button in the LCC would switch sequentially from caller to caller with each push. Identical black phones in the access portal were not part of the emergency net phone system.

Cableways

The control center and silo were connected to each other by means of steel tunnels called cableways. There were two cableways: the

▼ The unfinished long cableway and a portion of its internal framework can be seen in this photograph. Also note control center dome and air shafts (lower left), access portal at right, blast lock (center), and silo with airshaft and shock delay pipes (background). Note workers for scale.

USAF PHOTO, TMM ARCHIVES.

► This view of the long cableway looks toward the silo. Note accordion-like flexible joint (upper left and upper right). Also note internal suspended framework with telemetry cables connected to missile (left), electric power and control cables (top) and piping for chilled water, domestic water and compressed air (right).

short cableway that connected the launch control center to the blast lock via blast door 8 and the long cableway that connected the blast lock to the silo via blast door 9. Center to center, the distance from the LCC to the silo was 255 feet (78 m).

The short cableway was about 25 feet (8 m) long and the long cableway was about 140 feet (43 m) in length. Both were constructed of steel pipe sections 9½ feet (3 m) in diameter with walls a half-inch (1.2 cm) thick and were simply buried in the ground; they were not embedded in concrete. Both cableways were fitted with accordion-like flexible joints at each end, which permitted the tunnels to flex in response to shockwaves. A large inverted U-shaped framework independently suspended within the tunnels by spring canisters supported cables carrying electric power, control circuits and data signals to and from the silo and missile, and air conditioning water lines. Because the cableway and the U framework could move independently, the chances of a shockwave causing damage to critical electric and control cables was greatly reduced.

◄ Launch duct and missile engine emergency spray controls (center), standard fire alarm pull box (left) and launch duct light switch (right). Legend maintains that the spray controls were covered after several accidents involving fumbling for the light switch. These controls were located near the silo end of the long cableway.

► This indicator panel, located in the long cableway, displayed the status of silo blast valves, plug valves, work platforms and access doors.

1 Lt. Col. Chuck Smith, USAF (Ret), et al, interview

2 Ralph M.Parsons Company, Operational As-Built Blueprints, sheet 68.1

3 Stumpf, 118

4 Ralph M.Parsons Company, *First Construction Phase Structural Calculations,* 165

5 Lt. Col. Orville Doughty USAF (Ret), and David Stumpf, *Everything You Ever Wanted to Know about the Blast Locks but Were Afraid to Ask,* TMM training booklet, 1991, 2

6 Yvonne Morris, former AF Captain and MCCC, interview

7 OO-ALCP 144-1, A3-12

8 Daniel Ford, *The Button: The Pentagon's Strategic Command and Control System,* (Simon and Schuster, 1985),157-162

9 Tsipis, 135-136

10 Doughty and Stumpf, 3

11 TO.21M-LGM25C-1, 1-25

12 Ibid.

13 Ibid.

14 Facility Engineering Change, Real Property Installed Equipent, MCL-3303, TMM archives

USAF PHOTO, TMM ARCHIVES.

▲ Crew members on alert. Crew members wore white jump suits until about 1967 when "crew's blues" were introduced.

CHAPTER 5

LAUNCH CONTROL CENTER

The launch control center (LCC) was a three level structure. Level 1, the topmost level, contained the crew's living quarters, often described as a kind of "Motel 2" because of its austere accommodations. Level 2 was the control center with equipment needed to receive the launch order and launch the missile. The lowest level, level 3, was a support-equipment floor. Levels 1 and 2 were the only designated smoking areas in the underground portion of the complex.

Crew Members

A discussion of the LCC should begin with a review of the missile combat crew members and their responsibilities. Titan II crews consisted of four individuals: the Missile Combat Crew Commander (MCCC, said "M triple C"), the Deputy Missile Combat Crew Commander (DMCCC), the Ballistic Missile Analyst Technician (BMAT) and the Missile Facilities Technician (MFT). In most cases the duties of these four people overlapped. This was essential because, for example, if one crew person became unable to continue his or her duties, someone had to be able to fill in. Just as most systems on the complex had a backup, every duty needed a backup as well. What follows is a brief description of each crewmember's duties and responsibilities.

TMM ARCHIVES.

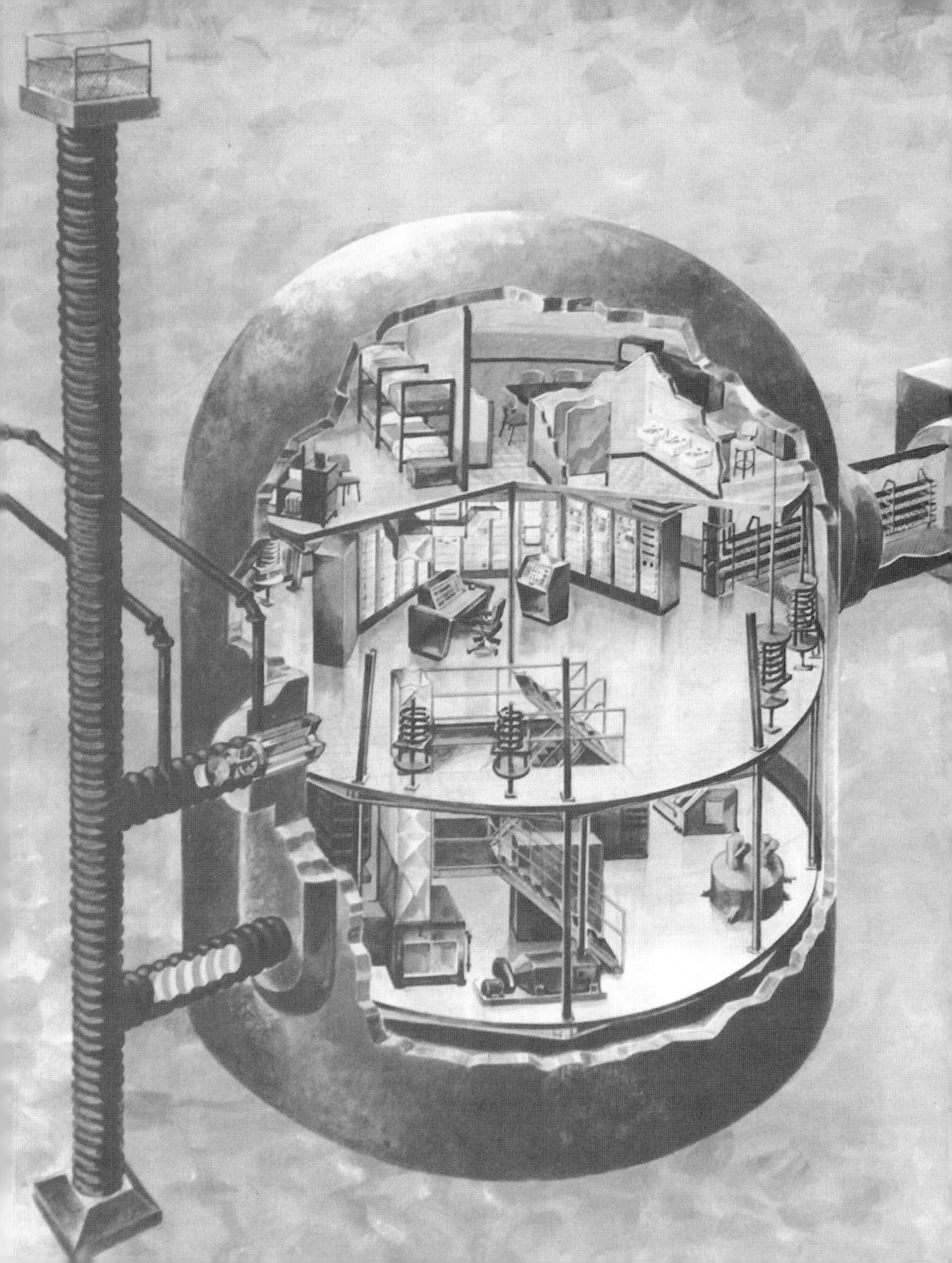

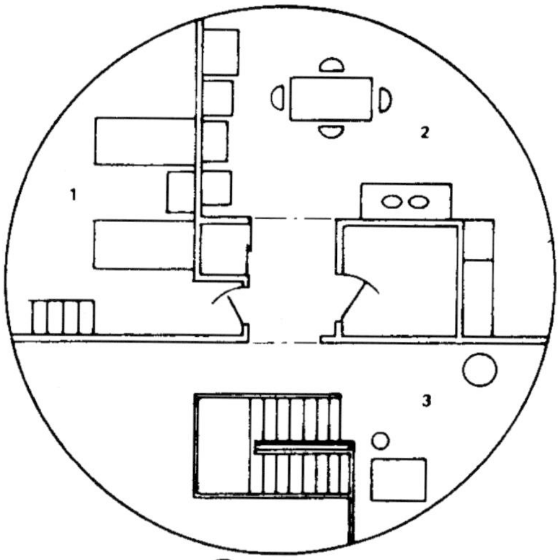

A LEVEL 1

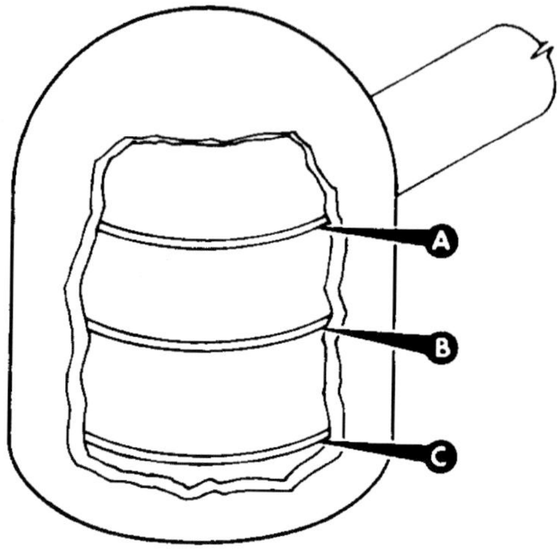

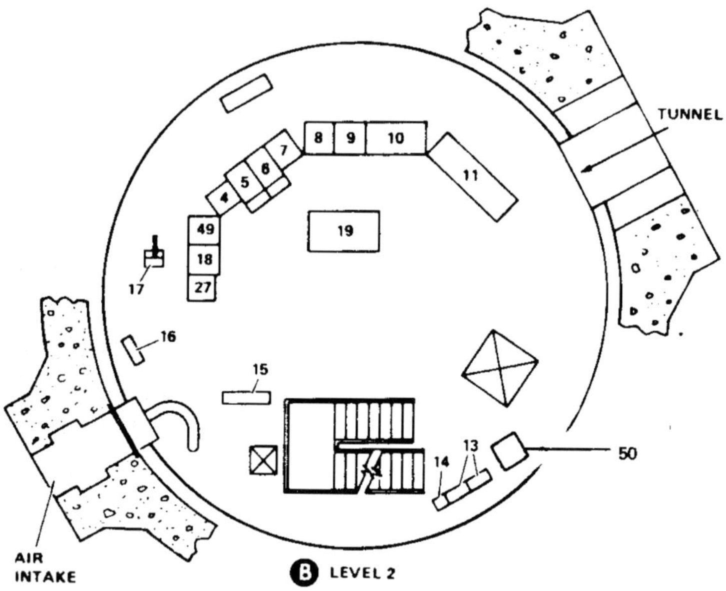

B LEVEL 2

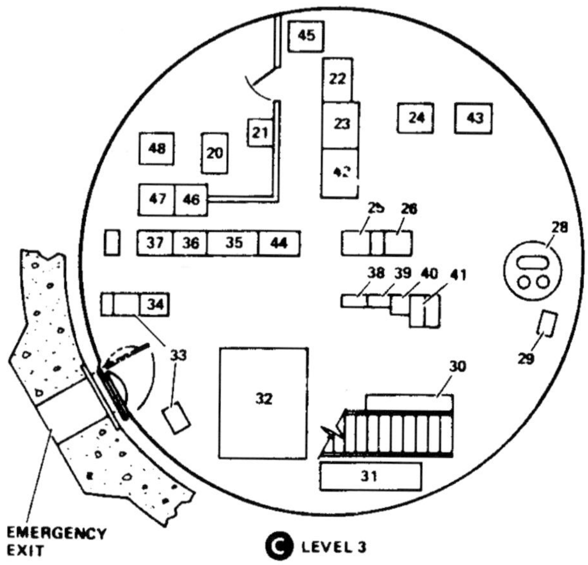

C LEVEL 3

1. SLEEPING QUARTERS
2. KITCHEN AND DINING AREA
3. STAIRWAY AND EQUIPMENT AREA
4. PRIMARY ALERT SYSTEM
5. ALTERNATE LAUNCH OFFICERS CONSOLE
6. VHF AND SSB RADIO
7. FACILITY POWER CONTROL BOARD
8. MISSILE SYSTEMS FAULT LOCATOR
9. CONTROL MONITOR GROUP
10. CONTROL POWER DISTRIBUTION
11. ALIGNMENT-CHECKOUT GROUP
12. (DELETED)
13. LIGHTING PANEL
14. CONTACTOR PANEL
15. HEAT AND VENTILATION CONTROL PANEL
16. DISTRIBUTION PANEL
17. HYDRAULIC HAND PUMP
18. SUBSCRIBER C
19. LAUNCH CONTROL COMPLEX FACILITIES PANEL
20. WIRE PROTECTION AND DISTRIBUTION CABINET
21. TELEPHONE CO. EQPT. CABINET (DMAFB ONLY)
22. MOTOR GENERATOR (DEACTIVATED)
23. BATTERY POWER SUPPLY 1
24. POWER SUPPLY 1
25. VOICE SIGNALING SYSTEM (VSS) CABINET
26. RADIO-TYPE MAINTENANCE NETWORK (RTMN) CABINET
27. HF-UHF RADIO
28. SANITARY LIFT
29. LIGHTING AND CONTROL TRANSFORMER
30. MOTOR CONTROL CENTER
31. AIR FILTER
32. A/C UNIT
33. FACILITY TRANSFORMER
34. CARRIER BAYS
35. COMMUNICATIONS 24-VOLT POWER EQUIPMENT BAY
36. WIRE CONTROL AND TRANSMISSION (WCT) EQUIPMENT BAY (ALOC)
37. WIRE CONTROL AND TRANSMISSION (WCT) EQUIPMENT BAY (LCCFC)
38. TERMINAL EQUIPMENT GROUP
39. MULTIPLEXER RECEIVER GROUP
40. MULTICOUPLER
41. TRANSMITTERS
42. BATTERY POWER SUPPLY 2
43. POWER SUPPLY 2
44. WIRE CONTROL AND TRANSMISSION (WCT) EQUIPMENT BAY (HV/HVPL SWITCHING SELECTIVE SIGNALING CIRCUITS)
45. ELECTRICAL INTERFERENCE FILTER GROUP
46. ELECTRICAL INTERFERENCE FILTER GROUP
47. ELECTRICAL INTERFERENCE FILTER GROUP (ACP ONLY)
48. ELECTRICAL INTERFERENCE FILTER GROUP
49. 487L
50. RTMN TRANSCEIVER BATTERY CHARGERS

TMM ARCHIVES.

▶ The crew's spartan sleeping room. The privacy screen was added when women were allowed as crew members beginning in the late 70s.

The MCCC, typically a captain or a major, was in charge of the complex. He or she (women were allowed to be crew members starting in the late 70s) authorized and coordinated all activities on the complex. The MCCC also copied, decoded, validated and authenticated all emergency action messages (EAMs). EAMs included everything from simple communication tests, to exercise messages, orders to change targets and the order to launch, among others.

The DMCCC, often a captain or first or second lieutenant, assisted the MCCC in performing his or her duties, and had the additional responsibility of monitoring communications to, from and within the complex. She or he also kept track of the location of all personnel on the facility, and like the MCCC, copied, decoded, validated and authenticated all EAMs.

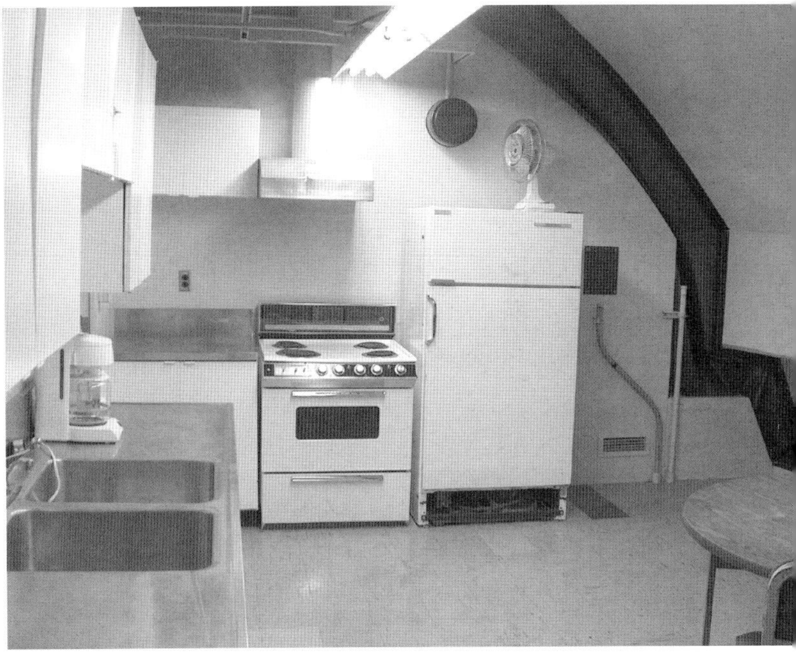

▶ A basic kitchen. People who could cook were highly prized as crew members.

The BMAT, an enlisted person, was responsible for monitoring the status of the missile and providing the MCCC with information concerning the missile's readiness condition.

The MFT, also an enlisted person, kept track of the status of the entire launch complex, and all the support systems therein, and provided the MCCC with information regarding the readiness of the complex.

The BMAT and the MFT also were trained to copy and decode EAMs.

LCC Level 1

Crew members worked a 24-hour shift known officially as an *alert*. Because a crew would be on alert for a full 24 hours, they needed a place to eat, sleep, relax, take a shower and so on. Level 1 of the LCC contained the crew's minimalist quarters. The bedroom contained two sets of bunk beds and a privacy screen for mixed crews. The kitchen was equipped with all the basics: a stove, refrigerator, sink, table and chairs. The restroom was fitted with a toilet, urinal, washbasin and shower. Also on level 1 were a hot water heater, a domestic water storage tank, and an air system exhaust fan.

A 250-gallon (950 l) storage tank held 124 gallons (470 l) of potable water. The extra volume permitted pressurization by the utility air compressor system on silo level 7.[1]

▶ This tank contained 124 gallons of potable water for the crew, many of whom refused to drink it. A water heater is partly visible at right

LCC Level 2

Level 2 of the LCC was the nerve center of the launch complex and was where the crew spent most of their time. It was here where the order to launch would have arrived, and from here where the crew would have launched the missile. From here the crew could monitor the entire complex and all the activities therein.

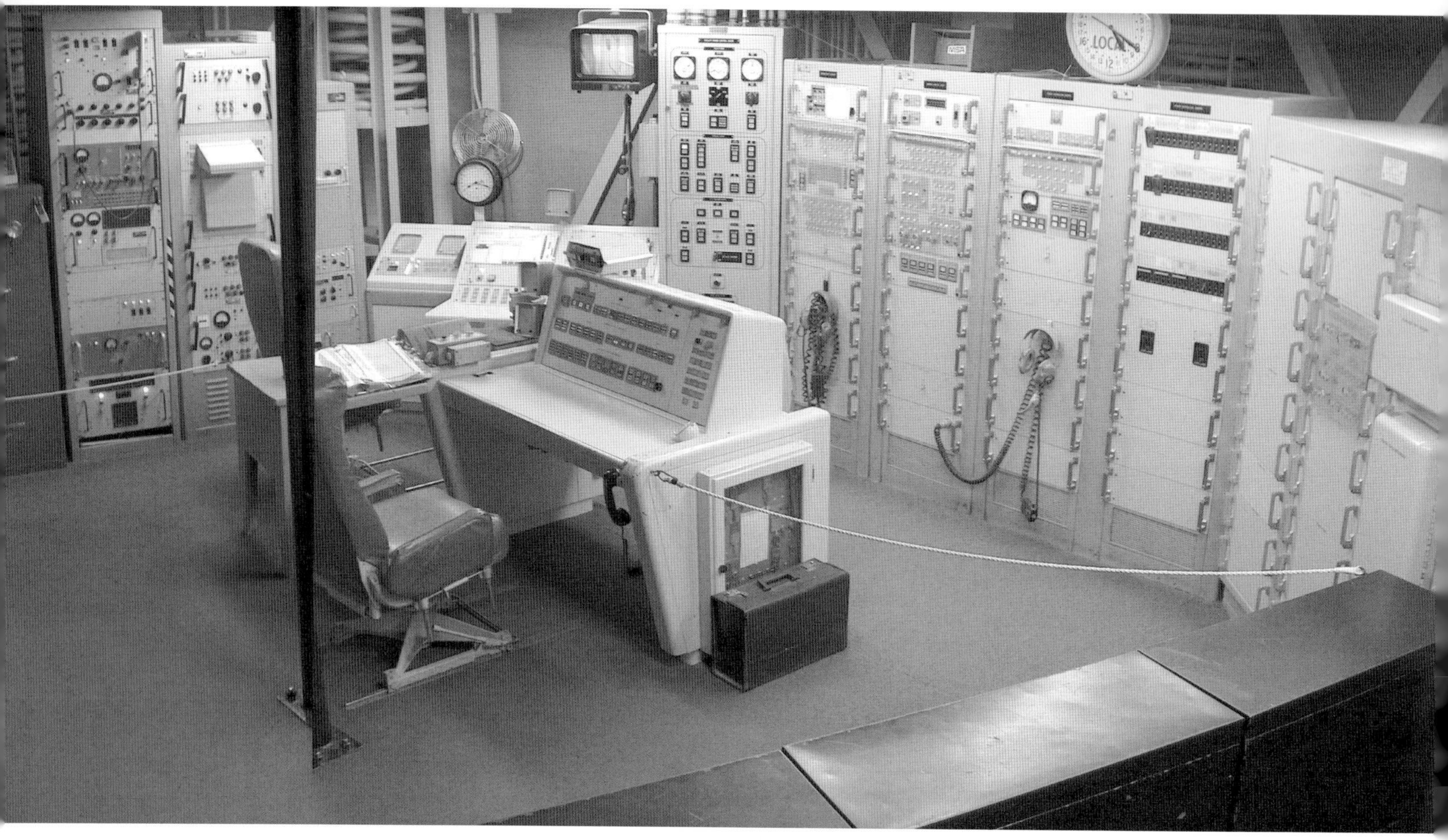

▲ The launch control center (LCC). The control complex facilities console (LCCFC) is at center.

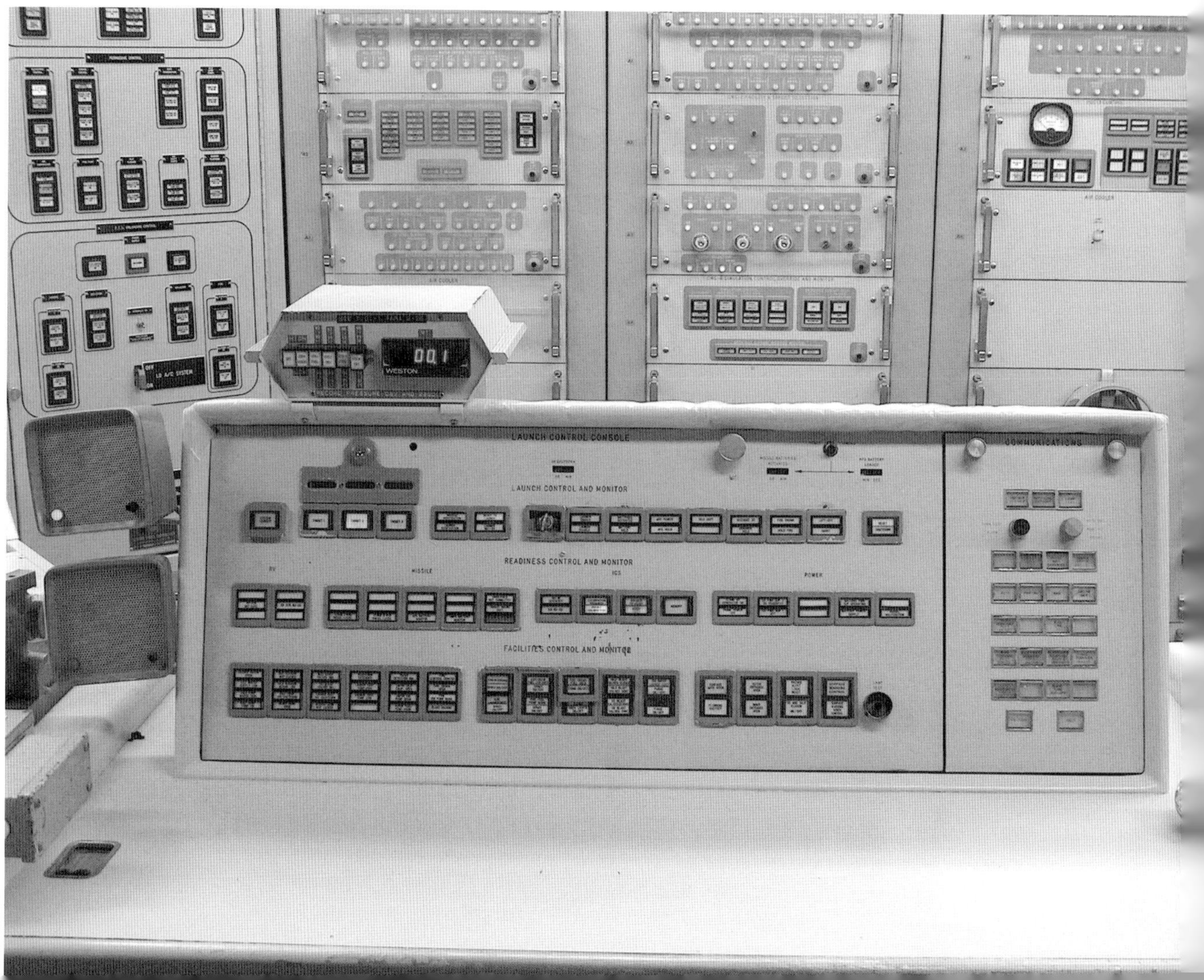

▶ The LCCFC. Think of it as the “dash board” of the complex.

▲ One of two key switches needed to launch the missile was located on the LCCFC. The small hole was for a lead security seal on the switch cover.

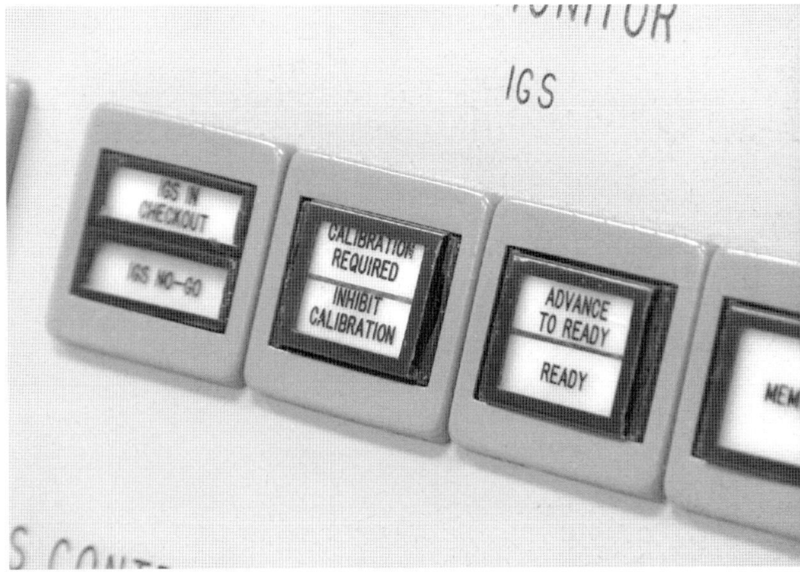

▲ These indicators displayed the status of the inertial guidance system (IGS).

Because the crew was responsible for a nuclear weapon, strict attention was paid to the Strategic Air Command (SAC) "Two Man Policy/No Lone Zone." While difficult to explain in a few words, the policy essentially dictated that at least two crew members were required to be present anywhere except LCC level 1. (On LCC level 2 at least one of them had to be a commissioned officer). In some cases it also meant that everyone had to be within eye contact of at least one other person. This "I watch you while you watch me" policy was designed to ensure security and also to guarantee that help was at hand if needed.

▲ Titan II carried only one warhead, but could be launched against any one of three targets.

Launch Control Complex Facilities Console

Level 2 was dominated by the launch control complex facilities console (LCCFC) in the center of the room and the racks of equipment that wrapped around it. The MCCC monitored the LCCFC, which provided an overview of the status of the missile and the complex. Its many indicator lights could be compared to the "idiot lights" on the dashboard of a car: they told the "mechanics" (the BMAT and MFT) where to begin looking for the trouble.

There were three rows of lights on the LCCFC. The top row related to the actual launch of the missile (and also used during tests) with indicators ranging from LAUNCH ENABLE to LIFT-OFF, and is where one of two launch keys would have been inserted if a launch had been required.

Target selection was determined by three target buttons on the left end of this row of lights. Changing targets was just a matter of pressing the correct button. Once pressed, the selected target button would light and a second or two later an indicator above the target selector button would also illuminate, confirming the new target.

All sites were assigned a primary target. Unless the crew received specific orders to change targets, the missile would be launched against the primary target. The targets were classified and were unknown to the crew, and remain classified as of this writing.

The middle row of lights on the LCCFC told the MCCC about the status of the missile, the reentry vehicle (RV), the inertial guidance system (IGS) and facility power systems.

Pushbuttons on the bottom row of the console controlled the surface warning lights and klaxon, the blast valves, and the main gate and entrapment area door locks. Here also were indicators that reported fires and controls that activated fire suppression systems. On the left

▼ Fire control systems and intrusion alarm.

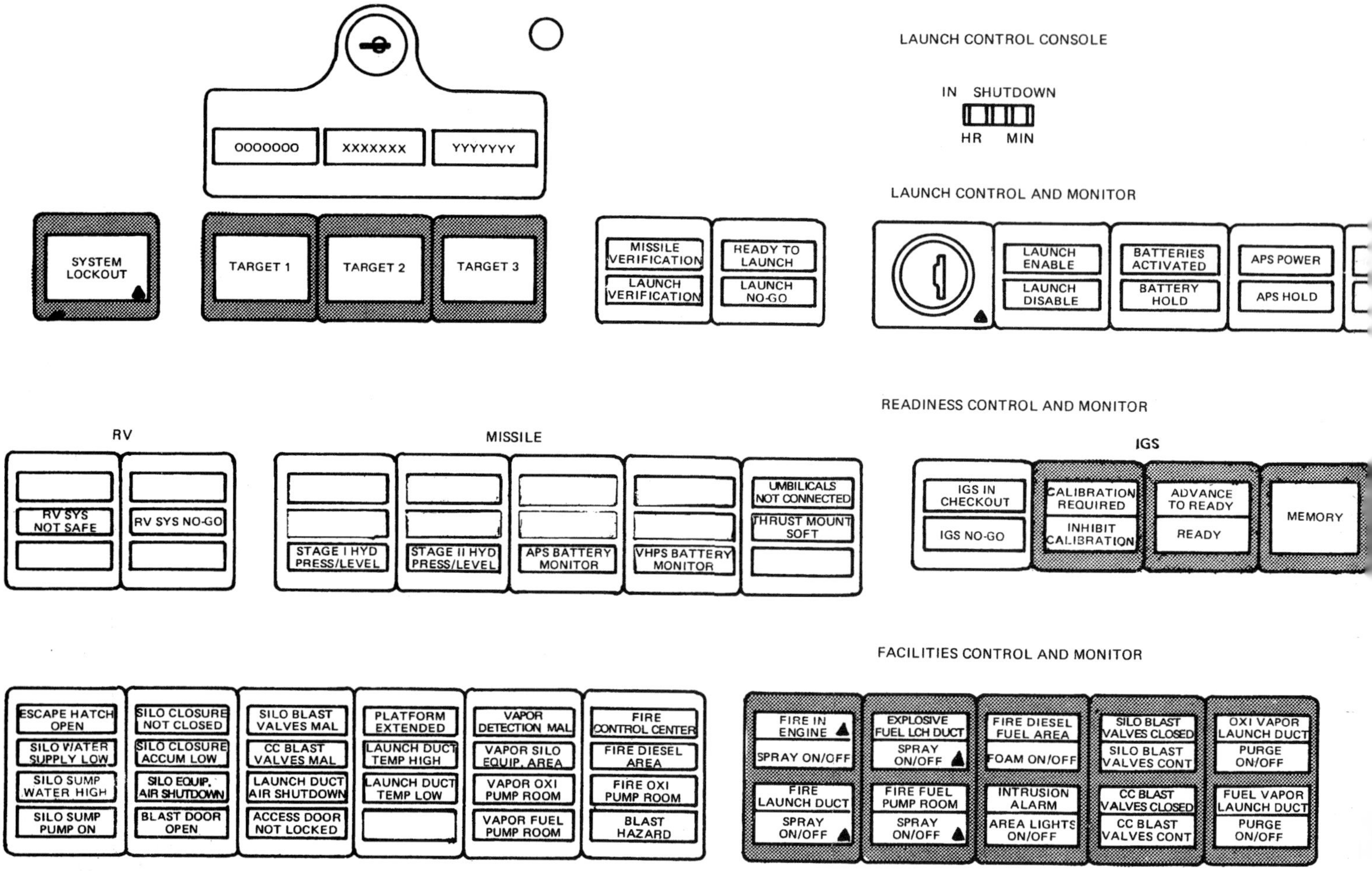

▲ There were a lot of things that could go wrong. These indicators gave the commander a "heads up."

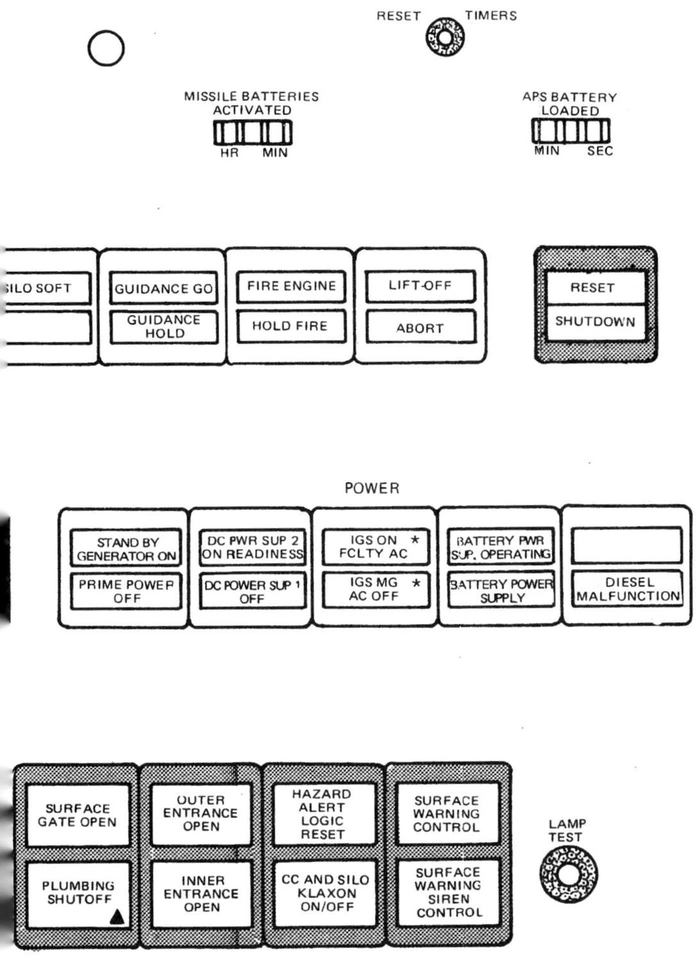

TMM ARCHIVES

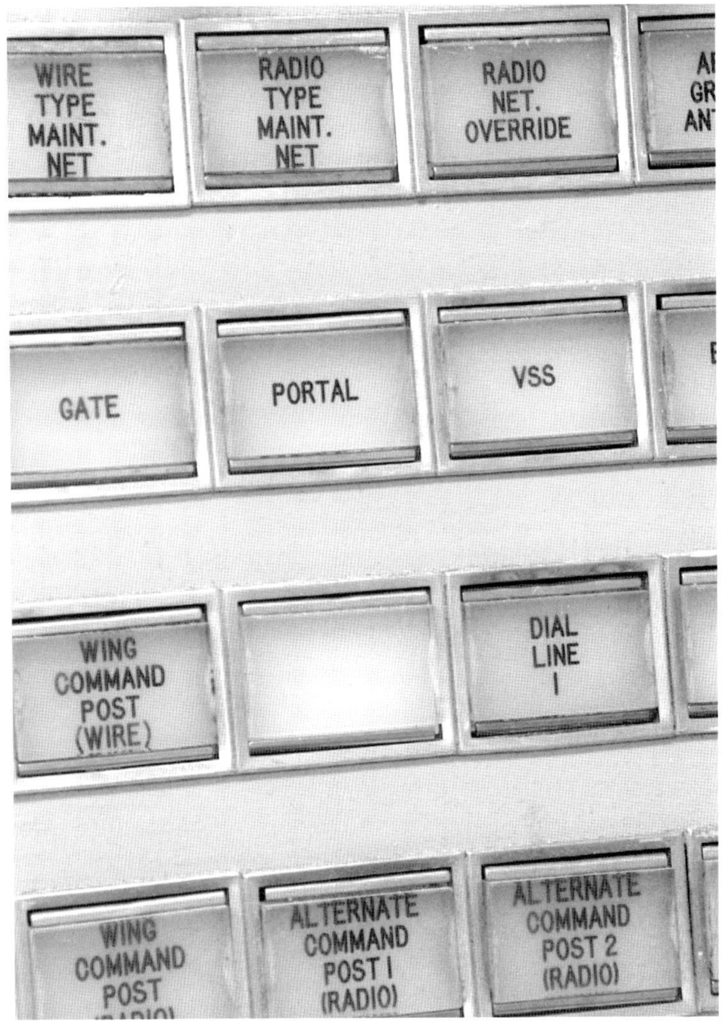

▲ Communications panel on LCCFC. Check with the wing command post, answer the gate phone, call home and find out what's for dinner.

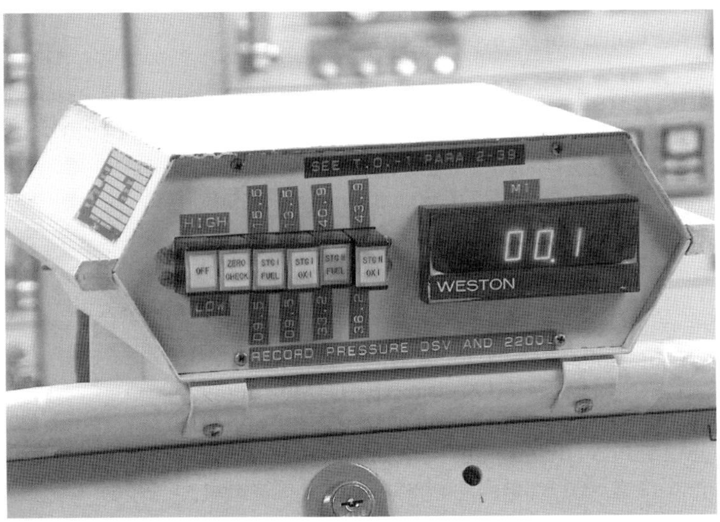

▲ A digital display provided a readout of pressure in each of the missile's four propellant tanks.

end of this row were additional indicators for a wide variety of potential problems in the launch duct and silo equipment areas. Indications such as SILO EQUIP AIR SHUTDOWN, VAPOR SILO EQUIP AREA or LAUNCH DUCT TEMP HIGH could signal the beginning of a bad day.

On the top left corner of the LCCFC was a small box with a digital readout called the propellant tank pressure monitoring unit (PTPMU). A series of buttons allowed the user to select which of the four missile tanks would be monitored. High or low fuel or oxidizer tank pressure was an indicator of something amiss, and would require swift action to avoid a more serious problem.

The PTPMU was added in 1976 to provide greater resolution of tank pressure than was afforded by the original HI-LOW pressure indicator lights on the LCCFC. Like the TEMP light on your car, it could be too late to do anything about the problem by the time the warning light turned on.

To the right of the LCCFC's main panel was a smaller communication control panel. From here the MCCC could activate the radio type maintenance net (RTMN) and wire type maintenance net (WTMN); talk with the wing command post, alternate command posts or other launch sites; make use of a standard phone line; and answer calls from the gate phone, phones in the access portal or emergency phones on the complex. Some communications systems required the user to push a button before talking and to release the button to listen. Anyone who has used a CB radio or walkie-talkie is familiar with this

▶ These were the speakers over which the launch order would have been heard. A set for the commander (foreground), and an identical set for the deputy (background).

▲ These controls unlocked blast door 6 and provided an indication whenever doors 6 or 9 were unlocked. Note built-in ashtray at lower right.

process, which is known as *push to talk* (PTT). So that the MCCC and DMCCC could have their hands free while using the radio, foot-switches on the floor beneath the LCCFC and ALOC controlled the PTT functions.

On the side of the LCCFC were two radio speakers. One speaker was connected to SAC headquarters near Omaha, Nebraska. For the 390th Strategic Missile Wing in Tucson, Arizona the other speaker was connected to the 15th Air Force (one of SAC's backup command posts) at March AFB in Riverside, California. It was over these speakers that the order to launch would have been heard. The speakers had only limited volume controls—they could be turned down a little, but not completely off.[2] At minimum volume they were still quite loud, perhaps because the LCC was, in general, a very noisy environment with its many fans and motors. Many crew members recall covering the speakers with gauze pads, Styrofoam, pillows or whatever else could be found to reduce the volume and the constant static that crackled from them.

Near these speakers were a black pushbutton and two large red lights. The button was one of two required to unlock blast door 6 (the other being located at the door itself). The lights provided a visual warning whenever doors 6 or 9 were unlocked. A small buzzer made sure the lights would not be overlooked.

Alternate Launch Officer's Console

The alternate launch officer's console (ALOC) was a sloped panel behind and to the left of the LCCFC. The ALOC was the responsibility of the DMCCC, and from it she or he had control of the HF and UHF radios, the RTMN and WTMN, the various antenna systems, and the telephone and VSS systems.

A pair of speakers identical to those on the LCCFC was located on the ALOC. Messages from higher headquarters were heard simultaneously on both sets of speakers. The crew received frequent messages, all of which ended with the command to "acknowledge now." Upon receipt of such a command, the DMCCC would press the "ack/call" (acknowledge call) button on the coded-switch status encoder panel. Doing so would turn on a light at the wing command post (WCP), indicating that the complex had received the message. The ack/call button would also have been used to let the WCP know that the crew had successfully launched their missile, had that situation ever come to pass.

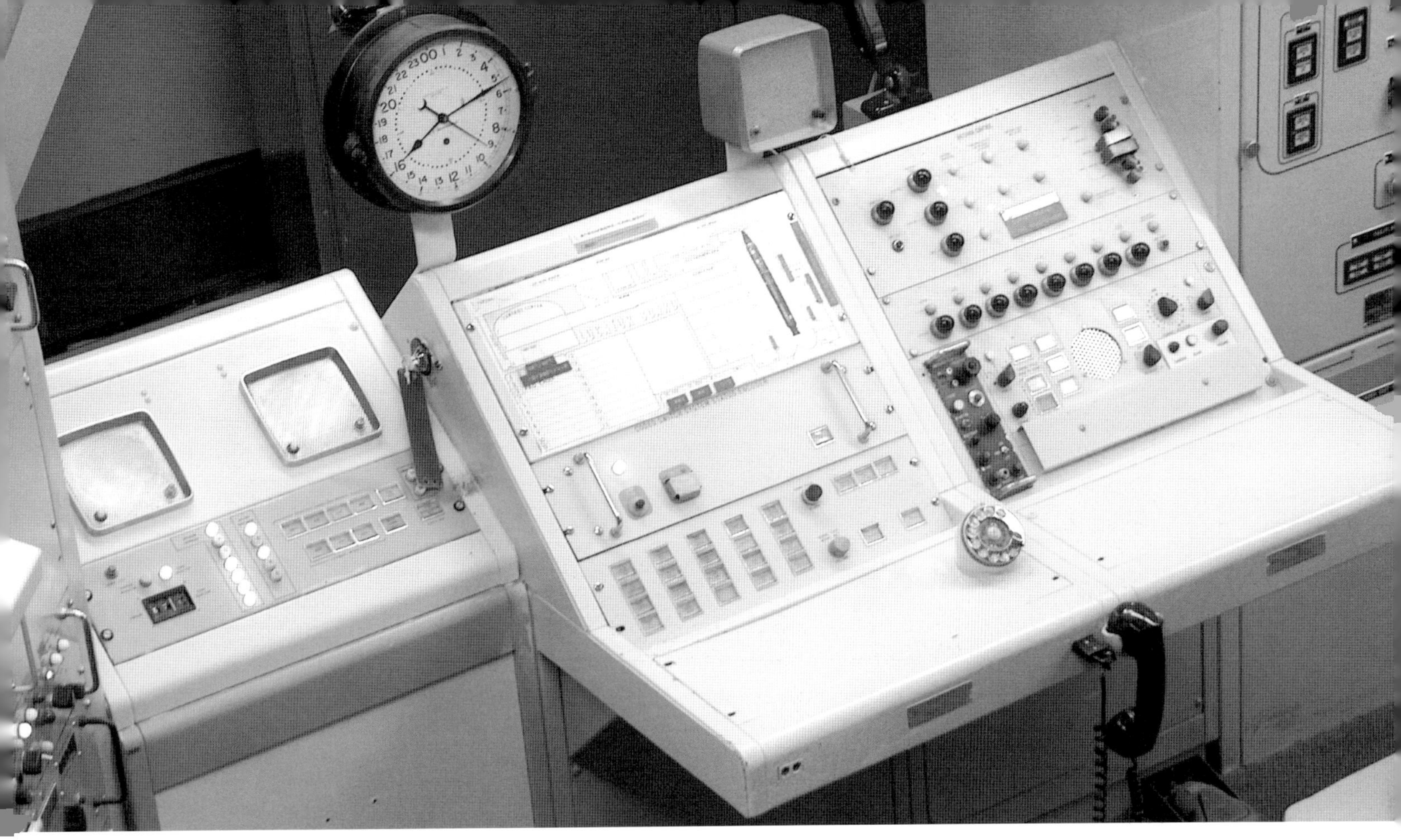

▲ The alternate launch officers console (ALOC) was the control panel for all the communication systems. Note launch key beneath clock.

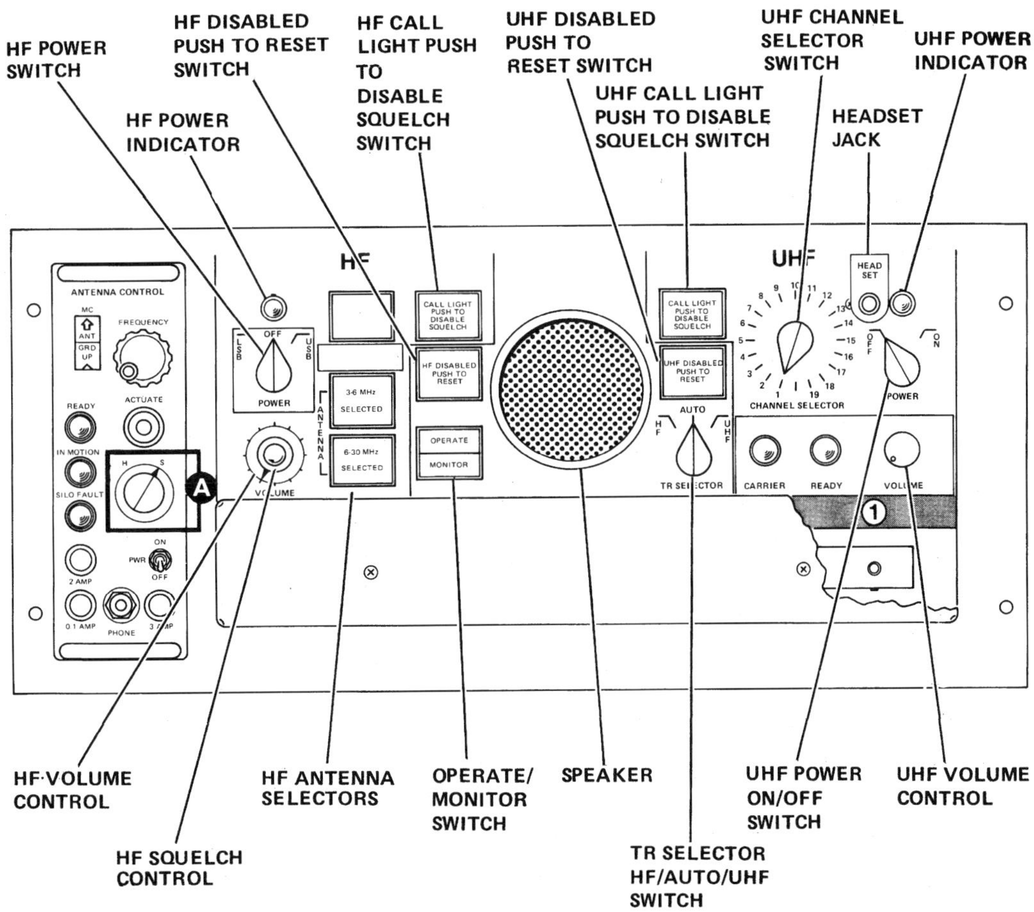

HF/UHF REMOTE CONTROL PANEL

TMM ARCHIVES

▶ HF and UHF radio panel (below). Hardened antenna controls and fault indicators (above). Note protective covers on pop-up antenna controls (upper right) to prevent accidental deployment.

▶ HF hard antenna tuning control. Dial in the desired frequency (top) and push the "actuate" button. The antenna telescoped up to the correct height.

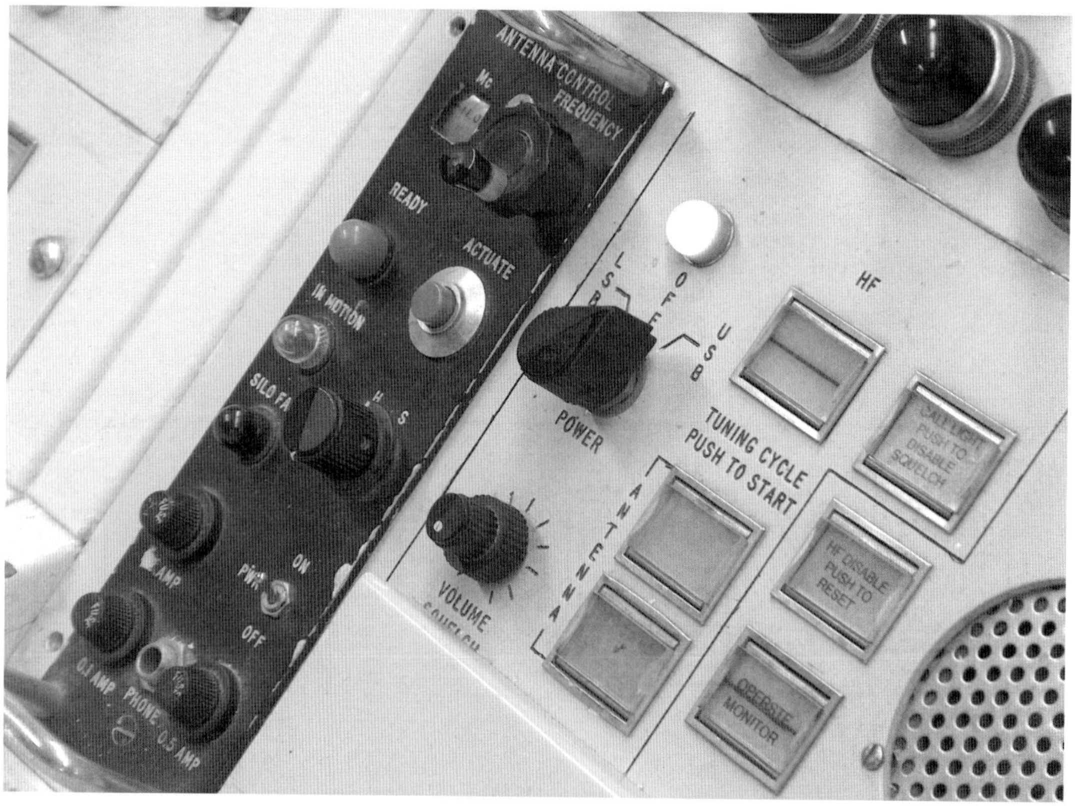

PHOTO: DWIGHT MEARS

The locator board on the ALOC was a graphic representation of the complex. Using a grease pencil (a forerunner of the dry-erase marker) the DMCCC could mark the location of personnel on the complex. This information would be critical in an emergency, and was generally helpful under any circumstance.

Also prominent on the ALOC was the emergency war orders (EWO) clock, also referred to as the launch clock. The launch clock told the official time on the complex. It differed from the standard electric clock (casually referred to by some as the lunch clock) in that it was "hacked" (calibrated) every six hours and was set to Coordinated Universal Time (for political reasons abbreviated as UTC),[3] also referred to variously as zulu, or Z time (see photo on page 60).[4] The advantage of UTC is that there are no time zones—UTC is the same time all over the world. This eliminated any confusion about what time was being referred to, and would prevent some wayward crew from launching their

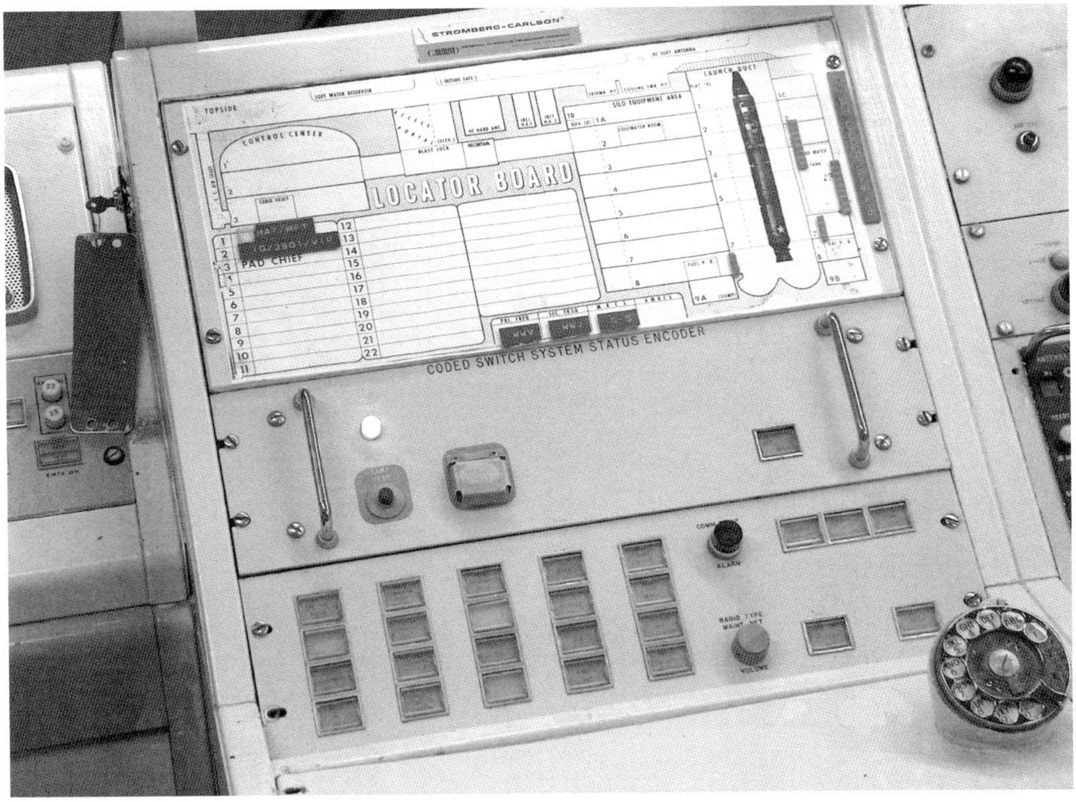

◀ Locator board (top), coded-switch status encoder (middle) with "ack/call" button at right, and communication panel (bottom). Note old-style rotary telephone dial (lower right).

missile an hour late, for example. The EWO clock was spring-wound so that power outages could not disturb its accuracy, and it was wound every Sunday morning.

It is worth noting that UTC is often spoken of colloquially as Greenwich (GREN-ich) Mean Time (GMT). Even crew members referred to the time as GMT. For the casual user GMT and UTC are identical. But strictly speaking, UTC and GMT are very different means of time keeping. In simplest terms UTC is based on an atomic clock while GMT is based on the rotation of the Earth.[5]

Two key-operated switches were required to launch the missile. As noted earlier, one of those switches was located on the LCCFC. The

◀ Each missile squadron had an alternate command post (ACP) to take command if the wing command post was destroyed. ACPs had additional communication equipment. Note pads covering speakers at right of LCCFC.

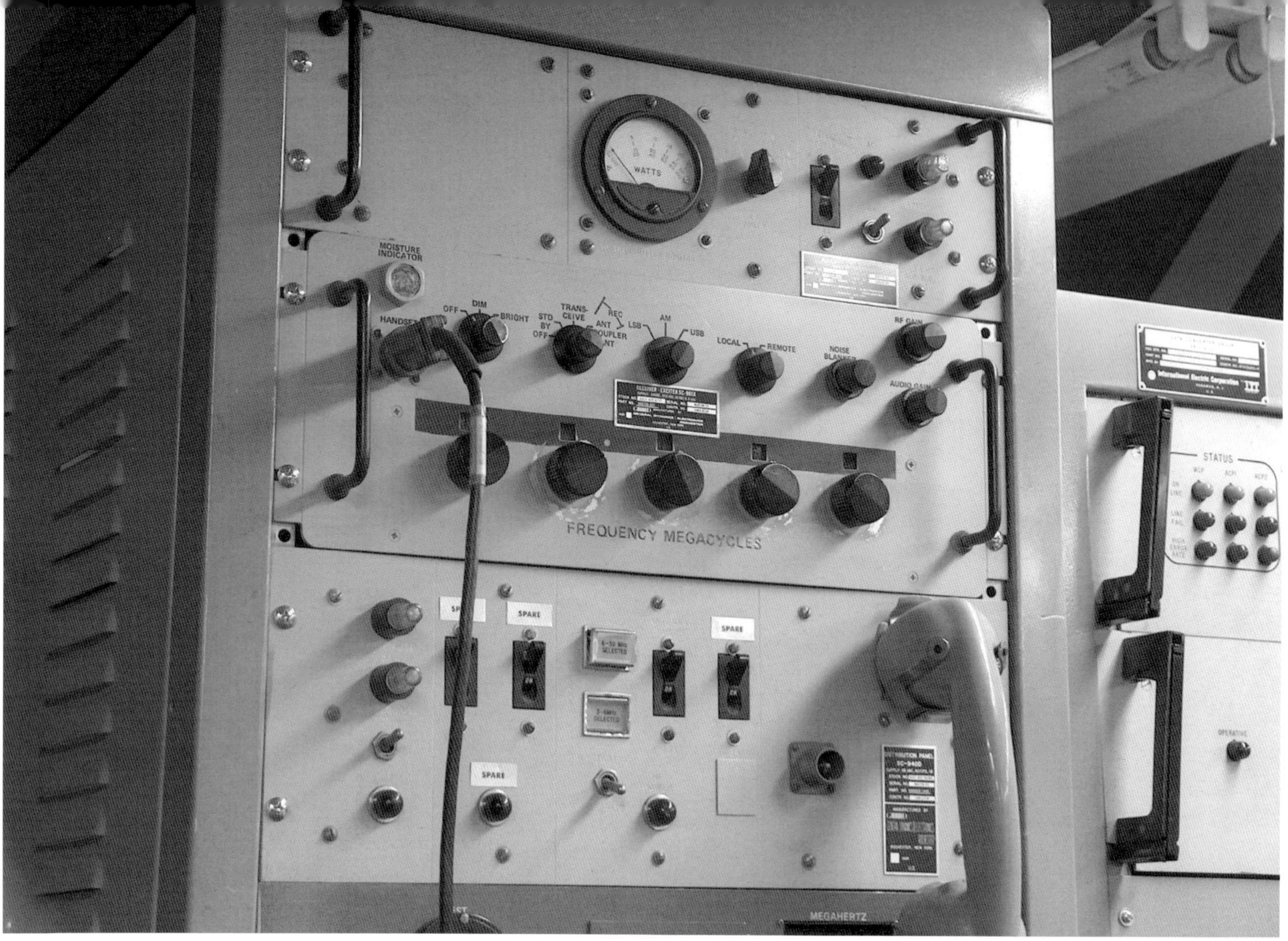

▲ HF radio unit made by General Dynamics. The radio had to be tuned to the desired frequency with controls on this panel, but could then be operated remotely from the ALOC. Note frequency readout in "megacycles."

other was located on the left side of the ALOC. (See photo on page 69.) Both of the keys needed to operate the switches were kept in a safe until actually required, and the keys switches were far enough apart that one person could not reach them both simultaneously.

Communications Equipment

Three large cabinets to the left of the LCCFC contained radio communications equipment. The leftmost of these racks contained the high frequency (HF) radio gear connected to the large discage antenna and to the discage's hardened telescoping backup. The site's HF radio transceiver (a combination transmitter and receiver) could operate anywhere between 3 and 30 megahertz (MHz), and provided single sideband (SSB) communications with a power of 1000 watts. SSB is a more power-efficient method of transmitting voice signals than amplitude modulation (AM), the kind used by AM radio broadcast stations, for example.[6] The HF radio provided worldwide communications capability. In the earlier years of the Titan II program the HF radio was used to communicate with what was called the "Bravo Net," a radio-gathering of military stations around the world and functioned as yet another backup system.[7] The HF radio was also used to listen to the National Bureau of Standards' (now the National Institutes for Standards and Technology) shortwave time station WWV, which transmitted precise time signals 24 hours a day. The crew used these signals to hack the launch clock.

Also in this equipment rack was the ultra-high frequency (UHF) radio connected to the small cone-shaped antenna located topside near the LCC air intake shaft. The UHF system was used mostly to talk to helicopters transporting maintenance or security teams or VIPs to the complex, but also to listen for the emergency rocket communication system (ERCS). The radio operated on any of 19 preset AM frequencies between 225 and 400 MHz.[8]

The center of the three communication cabinets was the 465L teleprinter equipment. 465L was a high-speed data network known as the Strategic Automated Command Control System (SACCS), the primary network for the transmission of emergency action messages (EAMs) to war-fighting commanders in the field.[9] The teleprinter provided printed copies of a variety of routine messages and copies of EAMs. Data for the 465L was received from the IRCS equipment located on LCC level 3. The 465L printer is often erroneously referred to as a fax machine.

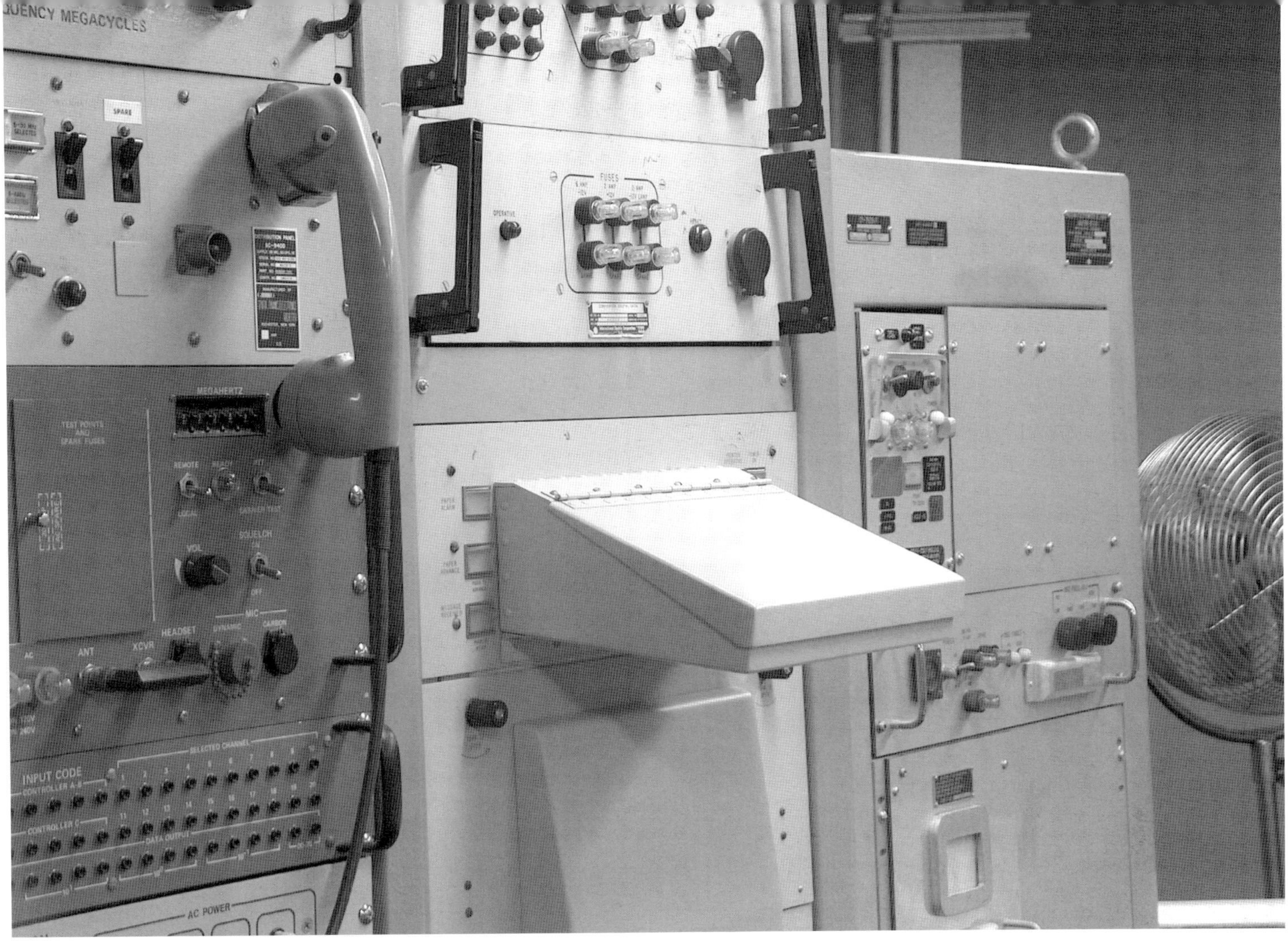

▲ The 465L teleprinter (center) provided hard copies of all emergency action messages.

In fact, the printer used the same dot-matrix technology employed in early home computer printers, but instead of a ribbon and ink it used a thermal process that essentially scorched characters on a special heat-sensitive paper.

The IRCS was a complex FM radio system carrying both voice and data signals for routine communications and EAMs, including the order to launch. The heart of the IRCS was a frequency diversity system operating on nine separate frequencies between 140 and 250 MHz.[10] Frequency diversity is the technique of sending the same message simultaneously on two different frequencies and letting each complex choose the frequency with the best quality. Frequency diversity is used to compensate for signal fading and interference. The process of choosing which frequency to listen to was accomplished automatically by electronic equipment within the IRCS. Eight of the nine IRCS frequencies were divided up with several subcarriers. Subcarriers made it possible for a single frequency to carry several kinds of information such as voice, data, and selective calling tones used to address specific sites, as well as system status tones used to automatically turn on backup frequencies in the event of a malfunction.[11] The IRCS automatically separated voice and data signals and routed them to the appropriate equipment—either the speakers on the LCCFC and ALOC or the 465L printer.

The IRCS voice channels provided communications between the wing command post (WCP) and other launch sites. A special IRCS channel called the primary alerting system (PAS) provided communications from SAC headquarters. While the message to launch would be transmitted on all of the site's communications systems (except, perhaps, for the telephone), the PAS was, as its name suggests, the primary means of transmitting the launch order to the crew.

An equipment rack to the right of the 465L contained the survivable low frequency communications system (SLFCS) receiver. SLFCS, officially designated as 487L, printed messages on a narrow strip of paper viewed through a small illuminated window. The paper exited through a slot above the window, much like an adding machine. When printing, it sounded very much like someone typing with one finger on an old mechanical typewriter.

SLFCS operated on very low frequencies (VLF), between 14 and 60 kHz. For comparison, standard AM radio broadcast stations operate between 530 and 1700 kHz. There are two

▶ 487L receiver equipment made by Westinghouse. The printer is at center, left.

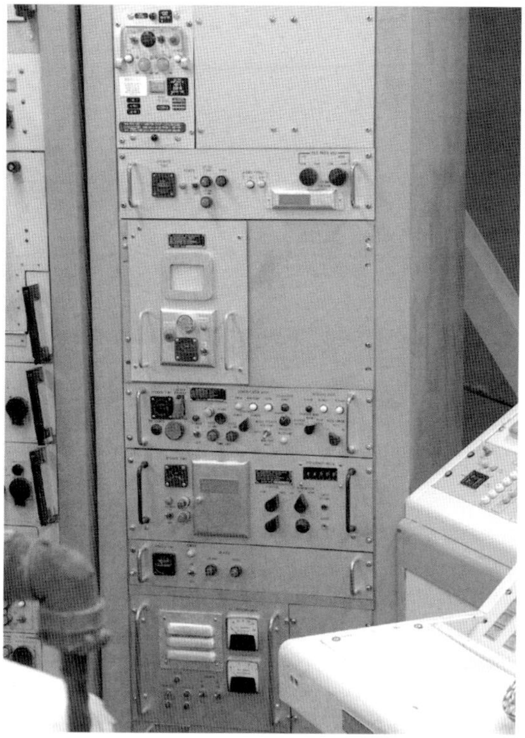

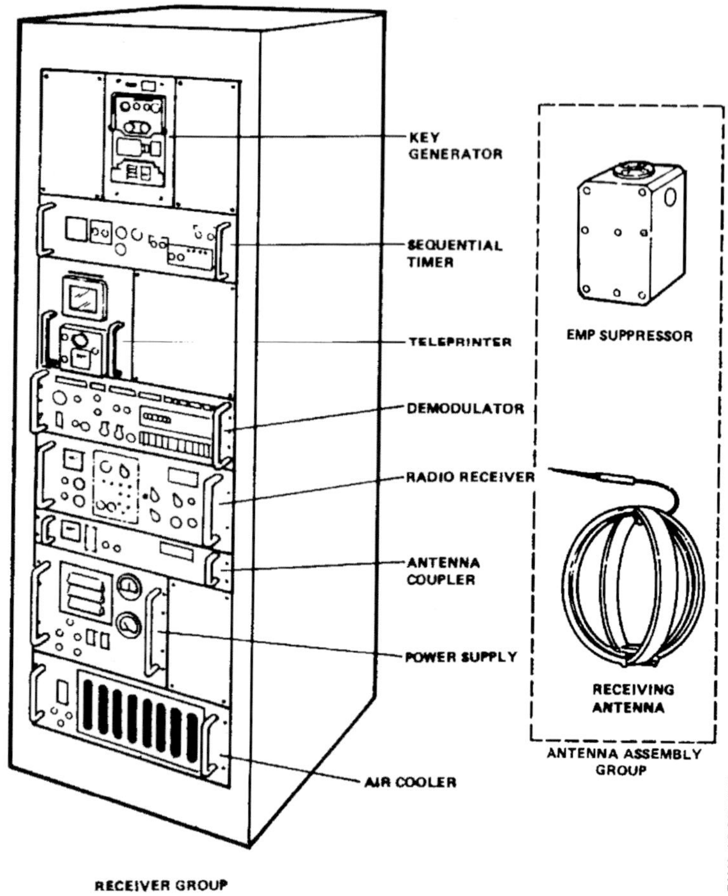

TMM ARCHIVES

important reasons for using VLF. One is that it permits the antenna to be buried, providing immunity from the heat and blast effects of a nuclear bomb. The second is that radio signals at these frequencies are largely unperturbed by the atmospheric effects of a nuclear detonation. HF radio frequencies, like those used with the discage antenna, must be able to bounce or "skip" off the ionosphere to reach the far corners of the planet. The ionosphere can be disrupted by the effects of a nuclear weapon exploding at high altitude, effectively cutting off HF radio signals. VLF radio signals bounce off the very lowest regions of the ionosphere where disturbances from a nuclear detonation would be almost non-existent.[12]

SLFCS could not carry voice messages, only data. Radio systems at such low frequencies operate with very limited bandwidth, and voice signals are too wide to fit. For the same reason, the transmission speed was very slow.

Curiously, for all of the sophistication and backup of the various communication systems, only three would work without some source of AC power. Only the RTMN, WTMN and the telephone had battery backup power supplies. And, of those systems, only the telephone would let the crew talk to the wing command post, assuming the telephone line and/or telephone switching office had not been destroyed during an attack, and then only until the batteries went dead. (Phone lines to a basic launch complex (BLX) were not hardened, and phone lines at the alternate command post (ACP) sites were hardened only for a few miles). This means that if commercial power failed and the backup generator wouldn't start or broke down, the crew would have been virtually cut off from the outside world. Not even the critical IRCS would function.

◀ The facilities power control board (FPCB). Meters at top display the status of commercial or generator power.

▶ One of the FPCB electric power monitors.

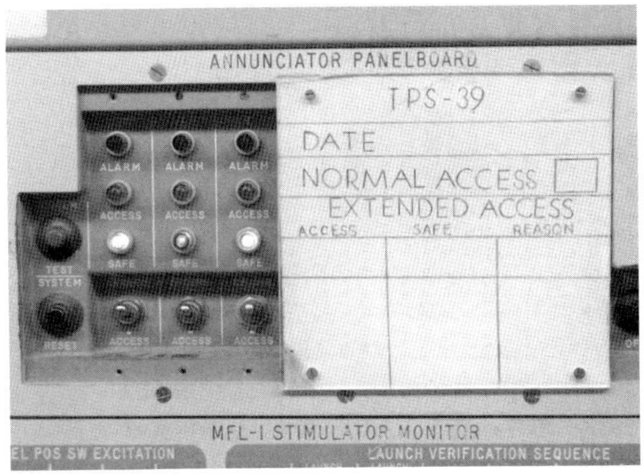

◀ AN/TPS-39 radar surveillance "tipsie" alarm panel. If an intrusion was detected a loud bell would sound and one of the "safe" lights would switch to "alarm."

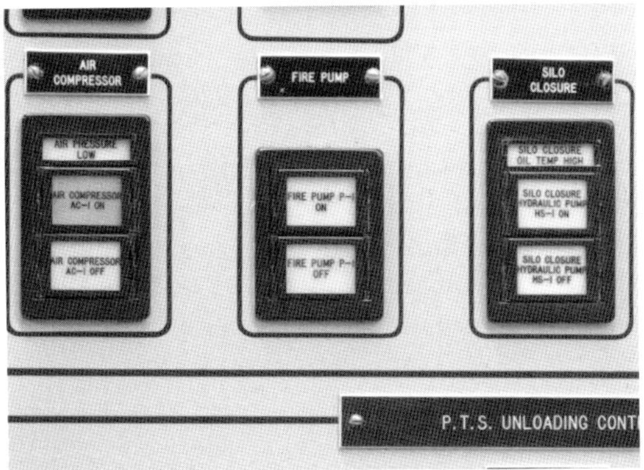

▶ A portion of the FPCB "permissive control" panel.

Facilities Power Control Board

Electric power was purchased from the local power company, but the complex was equipped with a backup generator. Three large meters on top of the facilities power control board (FPCB) indicated the voltage and frequency of the electricity supplied from the power company (or from the site's backup generator), and how much power (in kilowatts) was being used.

Lights on the FPCB indicated the status of numerous electrical systems. Controls permitted manual or automatic start of the emergency generator and provided permissive control of equipment such as the elevator, meaning that power to the elevator and certain other systems could be turned off or on from the FPCB.

Switches on the FPCB also provided for emergency control of propellant unloading equipment, though any unloading of propellant required manual connection of hoses to the missile. The exact nature of the emergency in which these controls would have been useful is not clear.[13]

Until about 1966 the bottom panel of the FPCB contained a Geiger counter to monitor radiation levels from the air intake shaft. This unit was removed and eventually replaced with a digital readout to monitor the temperature of

▼ Crew members perform a launch verification.

USAF PHOTO, TMM ARCHIVES.

A Few Words About Tech Data

Tech data, more properly called *technical orders* or "T.O.s," can be thought of as instruction manuals and pertained primarily to operating ground equipment (OGE), that is, equipment needed to keep the missile operating and ready to launch, and to the missile itself. The T.O.s contained diagrams, schematics, and operating and troubleshooting instructions for every system on the complex and missile, and there were hundreds of them. The most-used T.O.s were kept close at hand in a bookshelf in the LCC.

The mother of all T.O.s was referred to as the "Dash One." It was a single volume issued to all crew officers. The Dash One contained an overview of all systems and, more importantly, all of the essential checklists—including the launch checklist.

Another kind of technical order applied to real property installed equipment (RPIE). As mentioned earlier, this included all the heating, ventilating, air conditioning, plumbing and electrical systems, among others. The instruction manuals for RPIE were called the Strategic Air Command – Civil Engineering Manuals (SAC-CEMs, usually pronounced as "sack sim").

▲ The most-used T.O.s were kept close at hand.

the launch duct.[14] Once the Geiger counter was removed, the crew had no means of assessing the extent of post-attack radioactive fallout.

Missile Fault Locator

To the right of the FPCB was the missile fault locator (MFL) cabinet. The top panel actually had nothing to do with the MFL: it was the alarm panel for the AN/TPS-39 "tipsie" system. If the tipsies were tripped by an intruder this panel would sound off with a loud bell while a red light indicated which zone had been infringed.[15] The crew would then initiate the intrusion checklist and call the wing command post to request a security team. Under normal cicumstances crews would never go topside themselves to investigate. Three zones were defended by the tipsies: two around the silo door and one over the LCC air intake shaft.

▶ The missile fault locator unit tested both ground and airborne systems for proper operation. Tests were conducted as requested by the wing command post (WCP), SAC HQ, or as required by technical orders.

The MFL was located just below the tipsie alarm panel. The MFL was used during launch verification or missile verification procedures, which were conducted periodically to confirm that all missile and ground support systems were working properly and that a launch was possible if the need arose.[16] Launch or missile verifications were performed as ordered by higher headquarters, or as directed by what the crew called tech data.

▼ Some of the systems tested by the MFL.

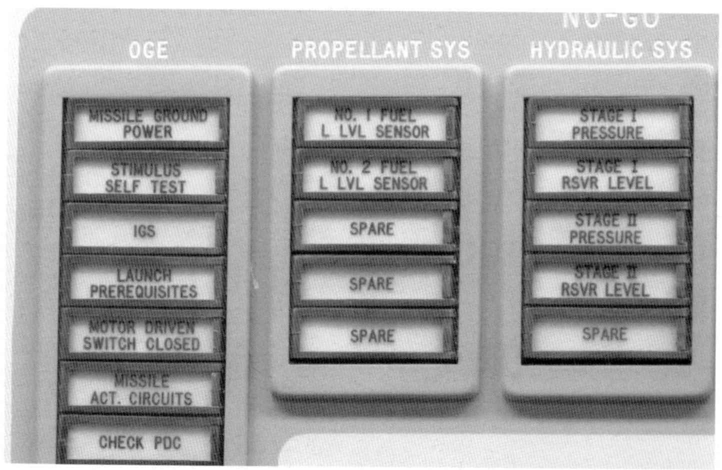

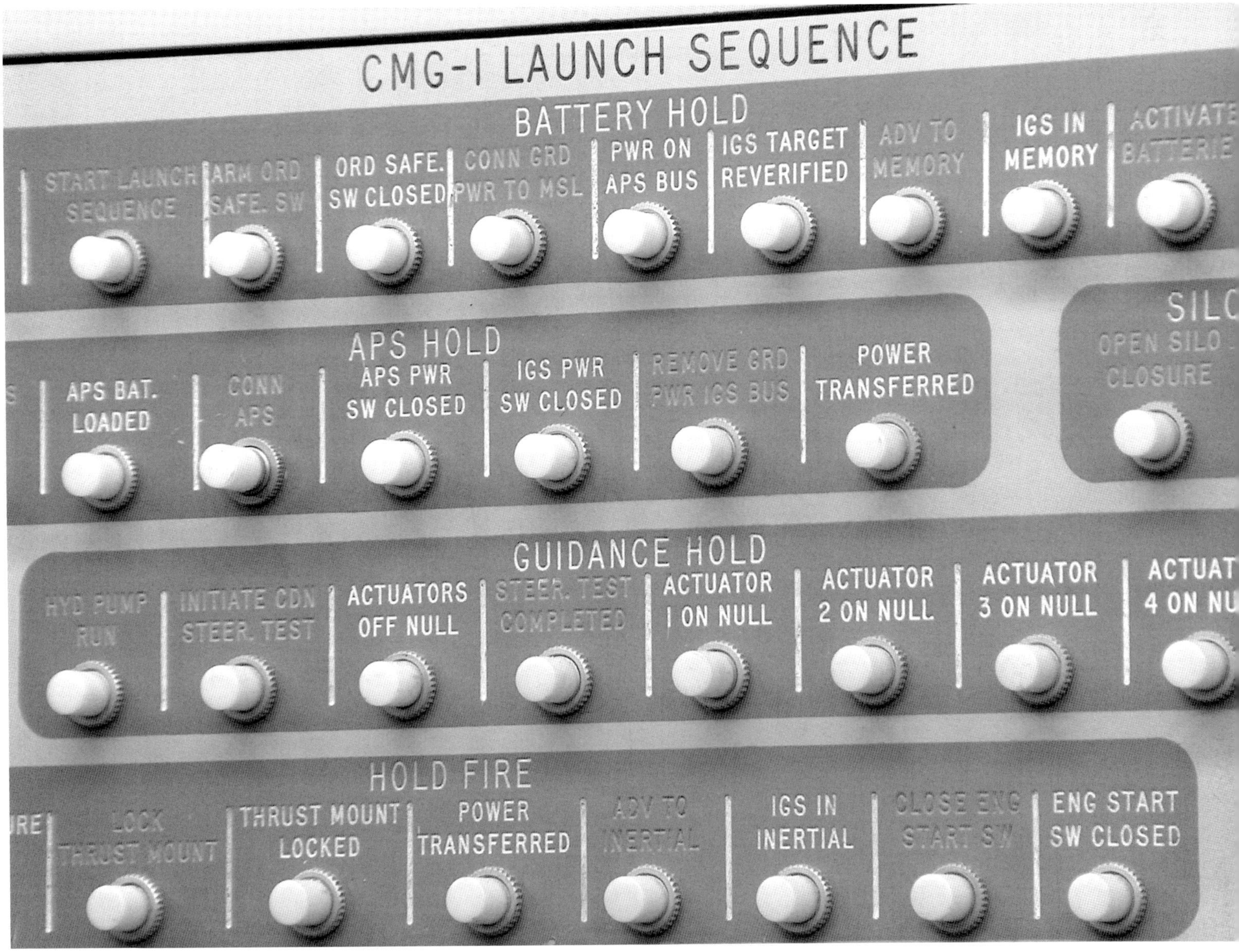

▲ CMG-I drove the launch sequence. Indicators with dark colored callouts sent commands to the missile and other systems (CLOSE ENG START SW[ITCH], for example). Those systems answered by illuminating the indicators with light colored callouts (ENG START SW[ITCH] CLOSED, for example).

► CMG units used dozens of circuit cards packed with thousands of germanium transistors and other discrete components, each soldered in place by hand. These units were designed at a time when integrated circuits were still 10 years in the future.

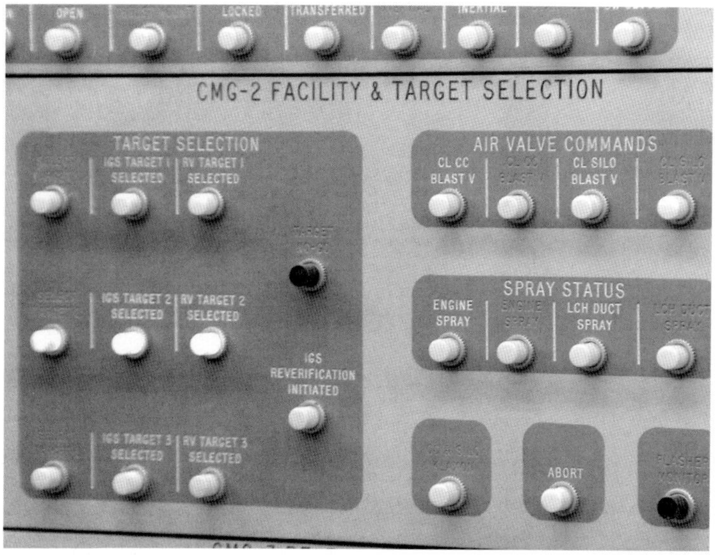

▲ CMG-2 confirmed the selected target and operated the silo blast valves and water sprays.

◀ CMG-3. Key operated switches were used to set how the warhead would explode—as an air burst or ground burst—upon arrival at the selected target.

Control Monitor Group

The control monitor group (CMG) was located to the right of the MFL cabinet and was divided into four sections. CMG-1 controlled the sequence of events during a launch. It sent commands to the missile and other systems, and received and interpreted signals from those systems to determine if they were functioning properly. CMG-1 could generate a "NO-GO" (a hold) in the launch sequence if problems were detected. By observing the lights on CMG-1, the BMAT could follow the progress of the launch sequence and troubleshoot problems that might occur.

CMG-2 was used to confirm that the selected target had been programmed into the guidance system. It also operated and confirmed the status of the LCC blast valve and the engine and launch duct water spray systems.

Information related to the re-entry vehicle (RV) and targets was displayed on CMG-3. This panel indicated the status of the RV and verified the selected target number and burst selection. Key-operated switches on CMG-3 permitted the selection of an air burst or a ground burst for each of the three targets. These switches were set when targets were programmed into the missile's guidance computer. The switch settings were not changed unless new targets were loaded. Targets were rarely changed and it is important to understand that the crew could not arbitrarily decide to change the settings of the burst or target selection switches.

▼ CMG-4, used in conjunction with the MFL, provided real or simulated operation of several critical systems.

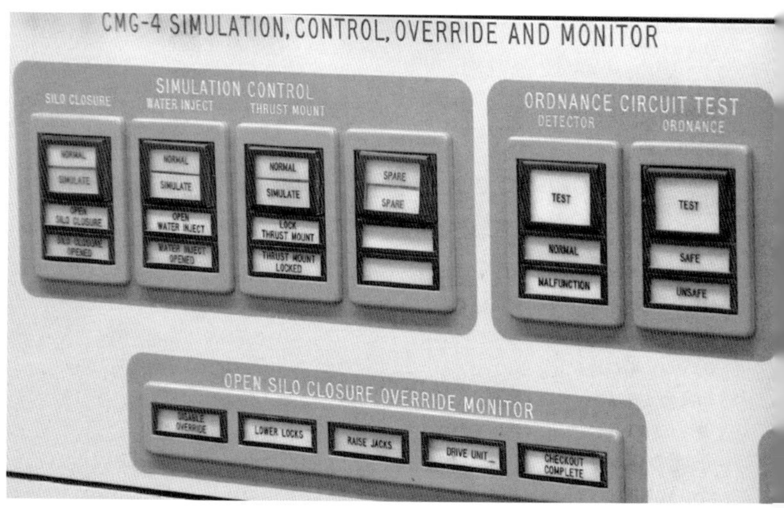

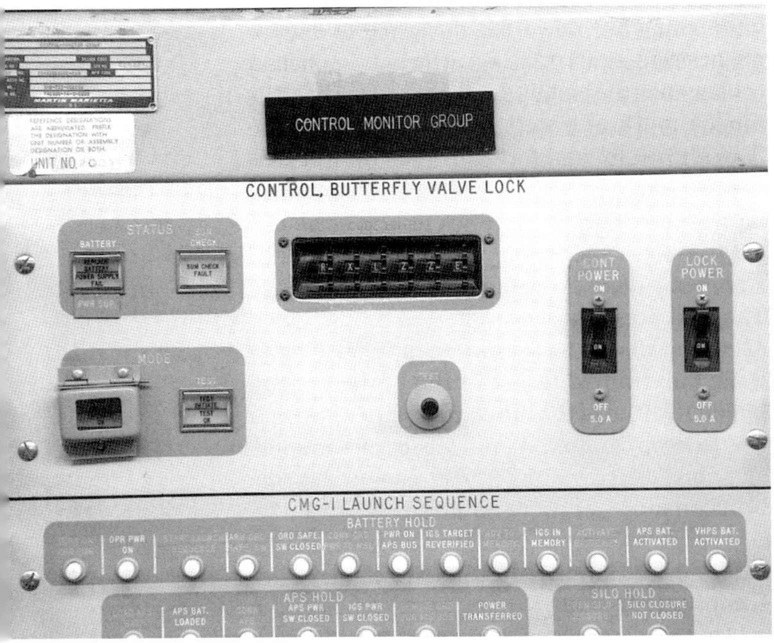

▲ The BVLC control. This combination lock prevented accidental, inadvertent or unauthorized launch of the missile. Its status was actively monitored at the WCP.

▲ The missile, and everything required to launch it, operated on 28 volts DC so that it could be launched with battery power alone, if needed. PDC-3 was used to monitor and control the 28-volt main and battery power supplies.

CMG-3 also permitted continuity checks of systems on the reentry vehical (RV) and could be used to check the warhead pressure. The warhead was pressurized with nitrogen gas to guarantee a stable internal environment for increased reliability.

CMG-4 was used in conjunction with the MFL in launch and missile verification procedures. CMG-4 could provide either real or simulated operation of the silo closure door as well as the sound suppression water deluge and thrust mount locking mechanism (discussed later).[17,18] Simulated operation of these systems was used when it would be inconvenient or unwise to actually operate them. For example, regulations prohibited opening the silo door for purposes other than an actual launch if rain, snow or other adverse weather conditions existed.

Butterfly Valve Lock Control

The butterfly valve lock control (BVLC) was a kind of combination lock that prevented unauthorized or accidental launch of the missile. The BVLC was installed in 1976 to replaced what was regarded as a less secure system. The unit was located in the top panel on the CMG cabinet. The BVLC used six thumbwheel selectors, each of which had 16 letters of the alphabet, to create a total of nearly 17 million possible combinations. The selectors had to be set to the proper six-letter code combination to unlock a valve in one of the stage-one engine oxidizer lines. Unless the valve was unlocked the missile could not be launched. A counter in the BVLC permitted only six attempts to enter the correct code before taking the missile off alert.[19] The BVLC was tested daily with special test codes that would not advance the counter or unlock the valve. The WCP was always notified before any test. The code for the BVLC was unknown to the crew and was not kept on the complex or even at the WCP. The code was revealed only as part of the order to launch.

On the missile, the valve and its electronics package were protected from attempts to disassemble or manually unlock it. A tamperproof housing included an explosive charge designed to permanently jam the valve in the closed position.[20,21]

The butterfly valve coded-switch control was the only component on the complex that was actively monitored at the WCP. Attempts to enter codes caused an indicator at the command post to illuminate. Crews never attempted to enter codes on the BVLC unless they had specific orders to do so.

Power Distribution and Control

To the right of the CMG/BVLC cabinet was the power distribution and control (PDC) panel. This panel was used to monitor and control the 28-volt main, and battery-backup power supplies. These units supplied readiness power (literally, power used to keep the missile ready to go), as well as transient power (power used to actually launch the missile). During a launch the PDC displayed the sequence of power-related events taking place on the missile. On the cabinet to the right of the PDC several rows of circuit breakers controlled a number of critical circuits in the silo and LCC. Normally all of the breakers were kept on except breaker CB-103, which controlled the ordnance power bus. The ordnance power bus supplied electricity to all of the explosive and pyrotechnic components on the missile—and there were dozens. They

▶ Banks of circuit breakers controlled critical circuits in the control center and silo. CB-103 controlled the ordnance power bus and was located second from left on the transient power distribution panel.

included the engine start cartridges, auxiliary power system (APS) battery squibs,[22] the hold-down bolts, staging bolts, vernier motors and pitch rockets, among others.[23] CB-103 was kept off to prevent accidental ignition of one or more of these devices. Accidental ignition of an ordnance device occurred twice in the history of Titan II: once prior to a test launch from Vandenberg AFB in California and earlier at a complex near Wichita, Kansas. In both cases the accidents happened when a short circuit in an igniter cable triggered ignition of a translation rocket on the missile. The translation rockets (discussed later) were small but powerful, designed to deliver about 5000 pounds (2300 kg) of thrust—sufficient to blow open the launch duct access door, bend its hinges, and hurl a maintenance person out the door with sufficient force to crack his hardhat against a shock isolation platform. There were injuries but miraculously no one was seriously hurt, and in each case the missile suffered significant damage but somehow stayed intact. The first accident was regarded as a random event, but when it happened again, investigators were determined to get to the bottom of the incidents. One of the outcomes of these events was to keep CB-103 off until just before launch, or unless required for a test. Eventually the translation rockets were removed.[24]

Missile Guidance-Alignment Checkout Group

The missile guidance alignment and checkout group (MGACG, usually pronounced ma-GACK) was used to program the missile's guidance computer. Targeting information stored on punched paper tape was fed into the MGACG and transmitted to the guidance computer aboard the missile. Note that target tapes contained only instructions on how to get to the target—the type of burst (air or ground) was determined

◀ The missile guidance alignment-checkout group (MGACG). Note technical orders (T.O.) bookcase.

▶ Delco's Universal Space Guidance System (USGS) control panel. The USGS replaced a dozen other panels and a dedicated motor-generator power source required by the old AC Spark Plug system.

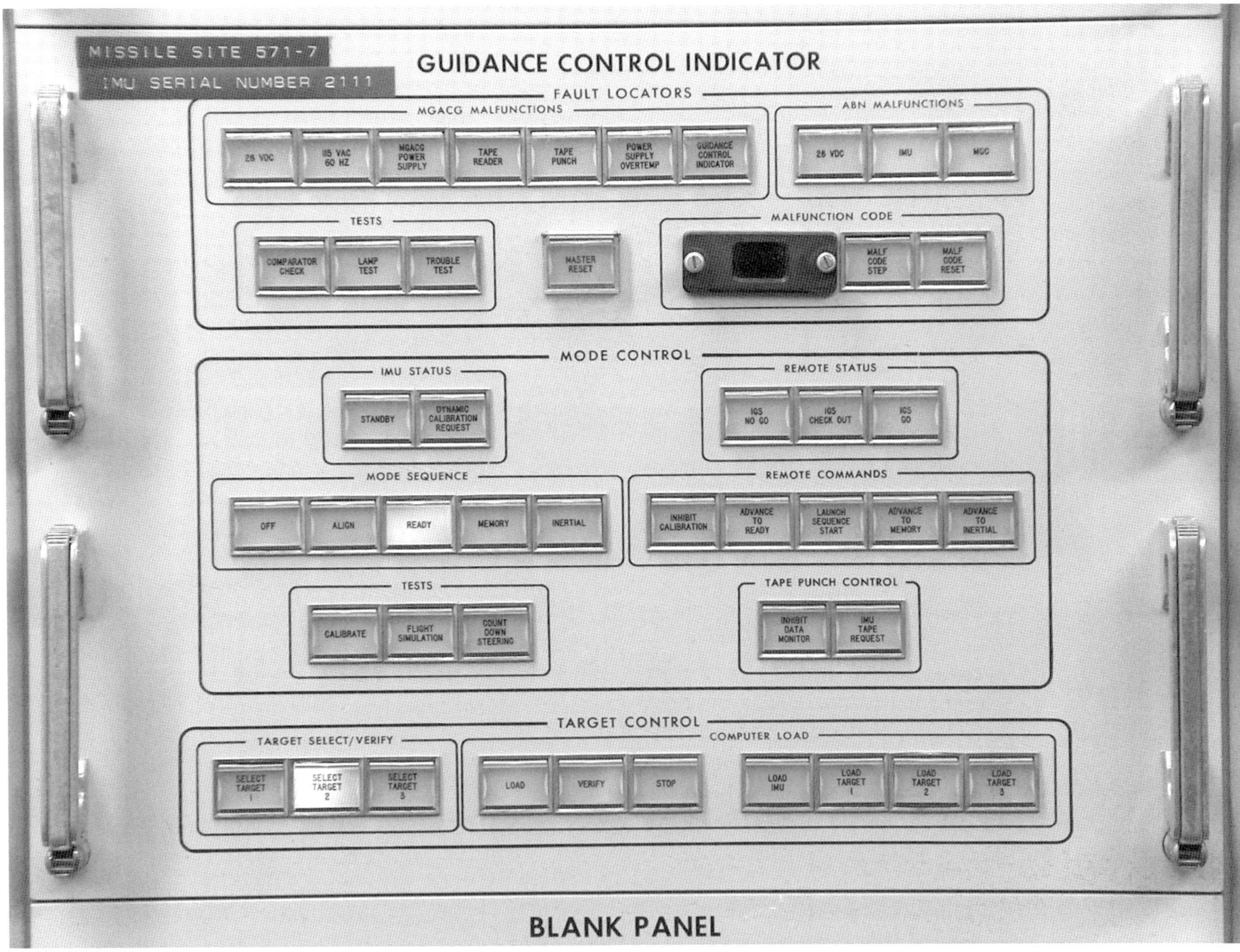

▶ Targeting information was stored on Mylar-backed punched paper tape. Though primitive by today's standards, punched paper tape will retain data decades longer than magnetic tapes or CDs. This tape is somewhat the worse for wear from 20 years of museum use, but probably would still work.

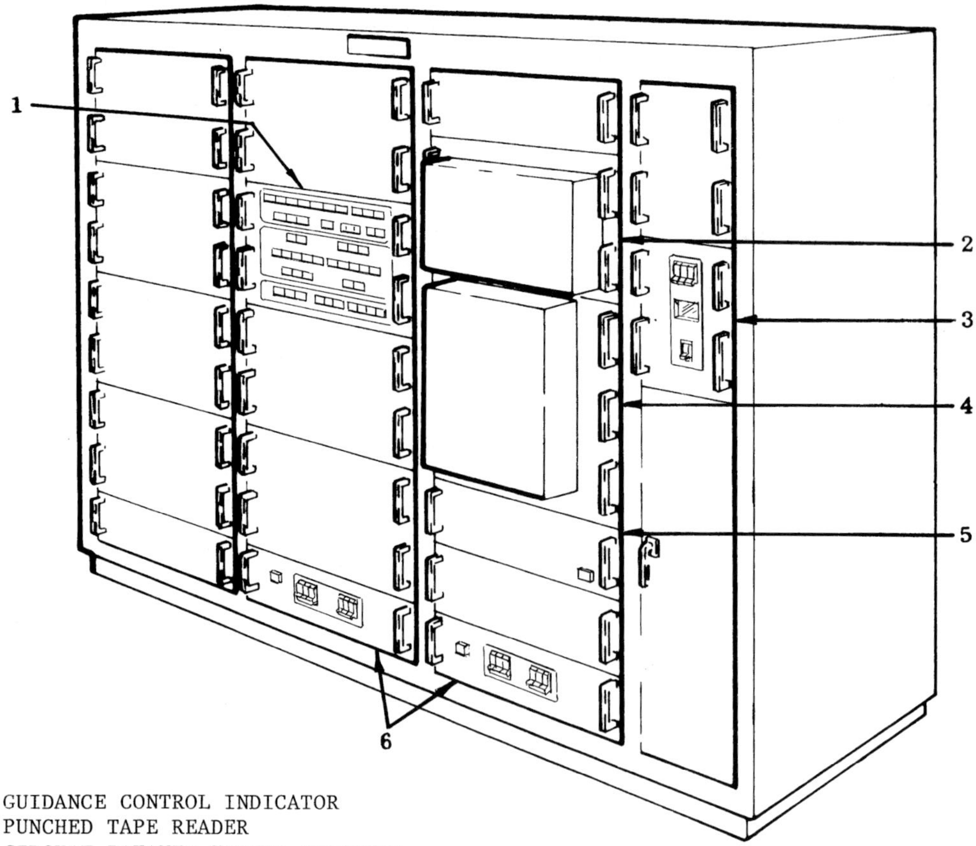

1. GUIDANCE CONTROL INDICATOR
2. PUNCHED TAPE READER
3. CIRCUIT BREAKER FILTER ASSEMBLY
4. SIGNAL DATA RECORDER
5. POWER SUPPLY
6. ELECTRONIC EQUIPMENT AIR COOLER

TMM ARCHIVES.

by the settings of the key-operated switches on CMG-3. The crew loaded targeting tapes when required, but as mentioned earlier, this was a rare occurrence. Target tapes were stored in the drawer beneath the red emergency war order (EWO) (pronounced EE-woe) safe (discussed shortly).

AC Spark Plug (a division of General Motors) made the original guidance system for Titan II in 1959, and until about 1978 the many panels on the MGACG now labeled BLANK PANEL were loaded with what was regarded (in 1959) as state-of-the-art electronics.

Titan II was originally planned to have a life of only ten years, but through a series of upgrades and modifications the program endured for more than 20. One of the consequences of this extension was that after ten years, spare parts for the guidance system began to run out, and many of the original manufacturers were either out of business or no longer willing or able to supply parts for the original system. Fortunately, a few years earlier, Delco Electronics had developed a new guidance system for Boeing's

▶ Air flowing down the intake shaft was routed around the blast valve piston by a lengthy "shock delay pipe." A blast wave would push the piston to the closed position. The valve would then be closed before the wave could travel the full length of the delay pipe.

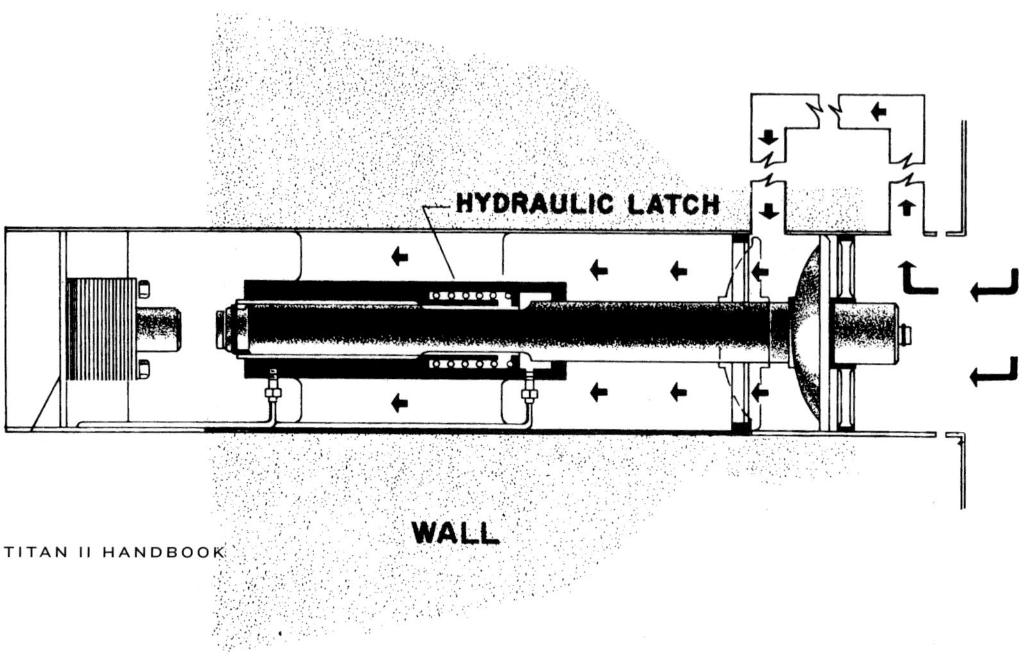

TMM ARCHIVES.

▶ The "go to war" safe contained authentication cards and the two keys needed to launch the missile. Lower drawers contained target tapes and a variety of classified documents.

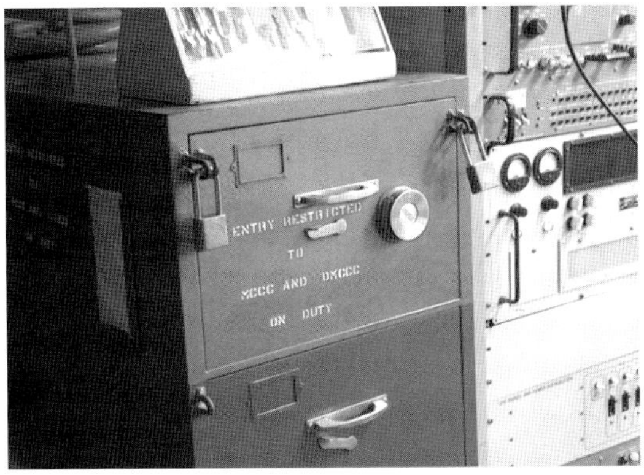

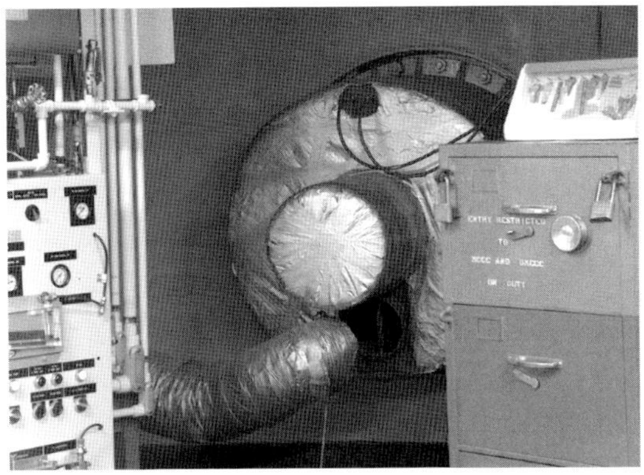

◀ This blast valve routed fresh air into the complex while protecting the crew from shock waves. It could operate automatically or manually, but once closed it had to be pumped open by hand.

747. It was later rebranded as the Universal Space Guidance System and used for the Titan III space launch program to deliver satellites into Earth orbit and later, to send probes to other planets. It was found that with very little modification, Titan III's guidance system could be used aboard Titan II.[25] Delco's system was so advanced compared to its 1959 predecessor that all of the equipment once needed to operate the guidance system was reduced to a single panel in the MGACG. Moreover, it replaced heavy and aging guidance equipment aboard the missile with state-of-art equipment weighing almost 50 percent less .

Emergency War Orders Safe

The red safe to the left of the communications cabinets was the emergency war orders (EWO) safe, often casually referred to by crew members as the "go to war" safe.

The EWO safe was locked with two combination locks. One of the locks belonged to the MCCC and the other to the DMCCC. The locks were the personal property of the officers and because the contents of the safe were classified the combinations of the locks were also classified. The two officers didn't know each other's lock combination, so both had to agree to open the safe. When a new crew arrived the officers of the off-going crew removed their locks and, after an inventory of the contents of the safe, the on-coming officers installed their own.[26]

The red portion of the safe contained message authentication cards and launch keys. Lower drawers contained target tapes and a variety of classified documents.

Air Intake and Blast Valve

On the wall just behind the EWO safe was the fresh air intake for the LCC. Behind the wall a three-foot (1 m) diameter air shaft extended all the way to the surface. Air was drawn down this shaft by a fan on LCC level 3. The difficulty with this system was that in a nearby nuclear detonation a shockwave could roar down the air shaft and demolish the LCC. To prevent this unfortunate circumstance the air shaft was fitted with a blast valve (BV-5), which would have been pushed closed by the force of a shockwave. The blast valve worked in conjunction with a shock delay pipe that gave the valve the time it needed to fully close. Closure of BV-5 protected the crew and automatically switched the LCC to re-circulated air. BV-5 would close automatically with as little as two pounds per square inch (psi) (0.14 kg/cm^2) of overpressure, and as part of the launch sequence (to preclude drawing missile exhaust into the LCC). The LCC blast valve could also be closed manually from the LCCFC during advanced states of readiness, or if the tipsie alarm sounded. Air from BV-5 was routed through a cleanable fiberglass filter on the valve

▼ Detail showing configuration of shock delay pipes.

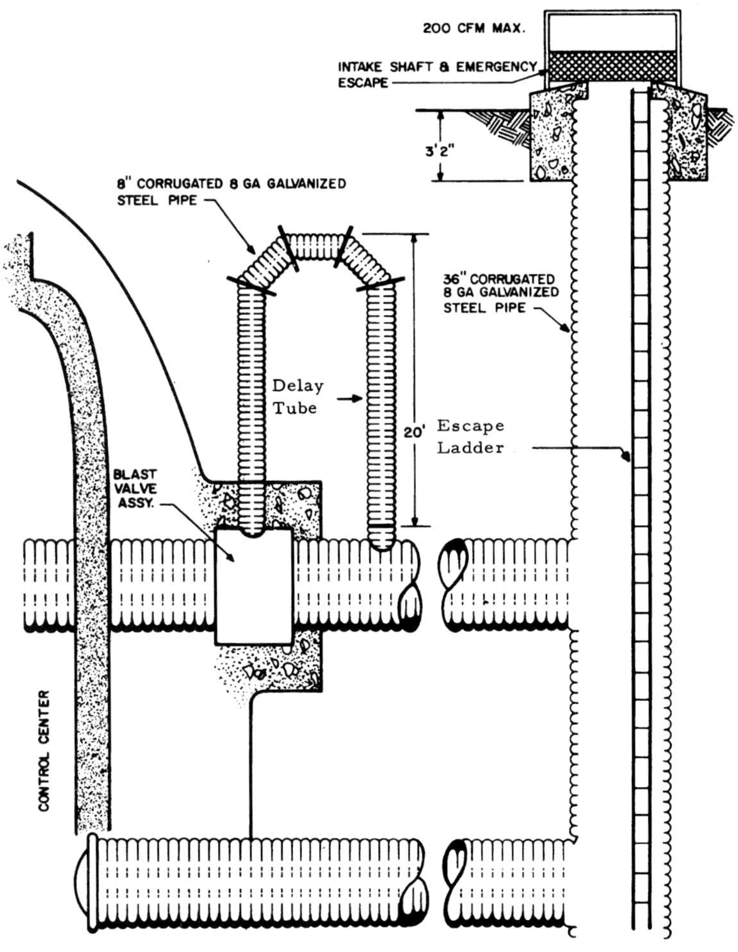

TMM ARCHIVES.

▸ HS-4 supplied hydraulic power to BV-5 and was pumped by hand. The unit was located behind the radio equipment cabinets.

◂ MCC-2 supplied power to the control center, blast lock and hard antenna systems, among others. Note CPR instructions at left.

itself, and then through a chemical, biological and radiation (CBR) filter located on the intake fan on level 3. This filter was characterized as a "fly paper" filter and could remove particles as small as one micron in size. It was not cleanable and had to be replaced when it became clogged.[27]

Hydraulic System 4

Hydraulic system 4 (HS-4) was located behind the communications racks. HS-4 powered BV-5 and had to be pressurized by hand—a lever on the front of the cabinet operated the hydraulic pump.

Weather Equipment

On the wall to the left as one entered the LCC was a panel containing chart recorders for the topside wind and delta-T temperature sensors. These units provided a record of wind and temperature conditions at the complex and were used to determine if acceptable weather

▸ Chart recorders displayed wind speed and direction (right), and delta-T temperature differential. Note Pad Chief's desk.

conditions existed for a propellant transfer operation to begin or continue.

Pad Chief's Desk

Below the weather recording equipment was a desk occupied by the Pad Chief. Not exactly a fifth crew member, the Pad Chief worked a more-or-less eight-hour day. He or she acted as a liaison between the crew and maintenance units at the base, kept track of routine maintenance issues, posted changes to the technical orders (T.O.s), serviced the diesel generator and scheduled any special maintenance that was required. The Pad Chief position was eliminated in the late 70s as part of a cost-cutting program, and the crew assumed those duties.[28]

LCC Level 3

The lower level of the LCC, level 3, housed the equipment needed to keep the control center operating.

Motor control center two (MCC-2) was the main power distribution center for the LCC. MCC-2 was located near the stairway from level 2 and contained a number of large levers that turned power off and on for all circuits in the LCC, blast lock and hardened antenna systems.

Adjacent to MCC-2, a long row of cabinets contained all of the equipment for the IRCS. There were two large 1000-watt FM radio transmitters called the F5 transmitters because they operated on frequency number five (hence the "F") of the IRCS. One transmitter was operational and the other was a backup. While the IRCS operated on nine frequencies, the crew could transmit only on F5. The remaining eight channels all received voice and data signals. All nine frequencies arrived at the complex on a single antenna. A special unit called the multi-coupler—a collection of cavity filters—was used to sort out the different frequencies used by the IRCS and direct each signal to the appropriate receiving and decoding equipment

▾ The IRSC equipment cabinets. General Electric 1000-watt F5 transmitters (left), multi-coupler (center) and multiplex receiver group (right).

▲ Cavity filters in the multicoupler unit. These filters separated the nine frequencies being received at the complex and routed them to the multiplex receiver group.

▲ Detail of IRCS multiplex receiver group plug-in modules.

in the multiplex receiver group. The multiplex receiver group was made up of dozens of plug-in receiver, decoder and oscillator modules occupying an additional rack next to the multi-coupler.[29]

Nearby, an even longer row of cabinets contained the electronics and switching equipment needed to operate and troubleshoot the telephone systems, VSS, RTMN and WTMN. Two nearly identical cabinets, wire control and transmission one and two (WCT-1 and WCT-2), contained all the connections, circuits and troubleshooting equipment related to the communications panels on the LCCFC and ALOC respectively.

The missile, and everything required

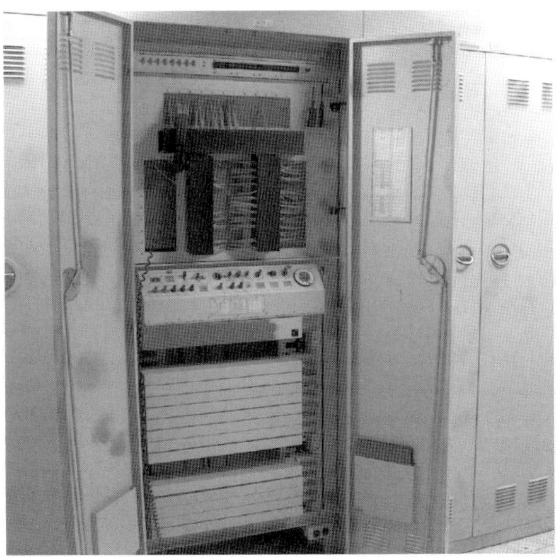

▶ The wire control and transmission cabinets were an interface between the LCCFC and ALOC, and all of the communications equipment and telephone lines. The narrow horizontal boxes in the lower section of the cabinet functioned as the site's miniature telephone switching office.

to launch it, operated on 28 volts DC. This permitted a launch on battery power alone, if worse came to worst. Level 3 housed a pair of 28-volt DC power supplies (PS-1 and PS-2) that powered much of the equipment on LCC level 2 and all of the equipment needed to launch the missile. If these power supplies failed, or if commercial power had been lost, electric power would have been provided by two large battery power supplies (BPS-1 and BPS-2) located adjacent to the main power supplies. These used nickel-cadmium batteries and were analogous to the uninterruptible power supplies used today in many home and business computer systems.[30] Exact figures on how long the batteries could have operated have not been found, but assuming a 50-ampere load, the batteries could have held up for only a few hours at most.[31] In the best case, the battery

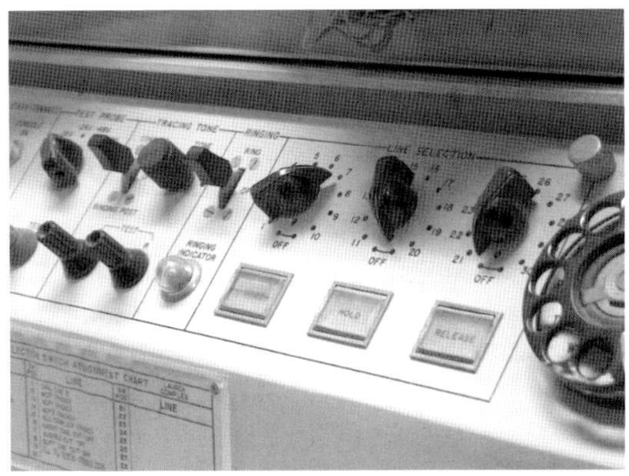

◀ Test equipment in the WCT cabinets permitted troubleshooting and re-routing of defective circuits.

▲ DC power supplies provided power for the missile and all the equipment required to launch. Conversion of power from AC to DC was accomplished with selenium rectifiers, the forerunners of today's modern silicon diodes. Selenium rectifiers were notorious for burning with toxic smoke when overloaded. The selenium units were replaced with silicon diodes when that technology became available in the early 70s.

▲ Battery power supplies took over if commercial power was lost or the main power supplies failed. These units used 210-amp/hour nickel-cadmium (NiCad) rechargeable batteries.

power supplies would have been required only for the minute or two it would have taken for the emergency generator to come on line.

Located between the 28-volt power supplies and the battery backup units was a motor-generator. The unit was powered by one of the power supplies (or battery supplies if normal power failed) and used a DC-powered electric motor to turn a generator to make AC power. It was used to power the MGACG prior to the installation of the Universal Space Guidance System around 1978. In those days, before high-power silicon diodes and other marvels of technology, motor-generators were the only way to change DC power into pure AC, and the only way to insure a totally uninterrupted source of AC power if the complex lost commercial power. Such a Rube Goldberg-like machine seems almost ridiculous today, when modern solid-state power inverters can make large quantities of AC power from an electronics package no larger than a shoebox. While the motor-generator would continue to provide power if commercial power was lost, it was not a "back-up" system *per se*. It ran the MGACG all the time. Curiously, this critical piece of equipment had no back-up of its own.

► The motor-generator ran portions of the old AC Spark Plug guidance system and was partially responsible for the hearing loss of many crew members. The unit was shut down and abandoned in 1978.

▲ 28 volt DC power supply control panel.

The motor-generator ran with a loud, monotonous high-pitched whine and was eventually equipped with spring mounts in an effort to hush its irritating din.

In fact, noise in the LCC turned into a serious issue. Crew members complained of temporary hearing loss due not only to the incessant buzz of the motor-generator, but also to the constant drone of the air conditioners, fans and blowers in equipment. Eventually the Air Force covered the tile floor with carpeting,

▲ AC-101 (at right) cooled the control center. This monster machine used cold water from the chillers on silo equipment level 2. The water was supplied via pipes running through the cableway.

and acoustic batting was hung in the area of the stairway leading up to level 1 and down to level 3. And with the installation of the new guidance system, the motor-generator was shut down and abandoned. These changes made a tremendous improvement, but one that came too late for many of the crew, a significant number of whom now need hearing aids.

Across the room from the motor-generator, and near the IRCS equipment was an air supply fan and air conditioner. The fan pulled fresh air into the complex through BV-5 at the rate of 200 cubic feet (5700 l) per minute, mixed it with recirculated air through the air conditioner, and then directed it into the LCC. Air was exhausted from the LCC by a fan on level 1 and was normally routed into ductwork and down the long cableway into the silo equipment area.[32]

Near the air conditioner was the emergency

escape hatch. A large, domed door could be unlocked with a ratchet-wrench and swung open to reveal a ten-foot (3m) tunnel that connected to the air intake shaft. The air shaft was equipped with a ladder that led up to a grating at the surface. At 571-7 the museum has installed an evaporative cooler on top of the air shaft, and the grating is no longer visible. Use of the escape hatch would have been necessary if the access portal had collapsed or if fire, smoke or toxic vapors prevented the crew from opening blast door 8.

Lastly, the wastewater ejection pump on level 3 operated as required to pump sewage from the complex to a septic tank located near the surface. In a condition of advanced readiness, the septic tank plumbing line would have been shut off and sewage would have collected in a 500-gallon (1900 l) ejection pump reservoir.[33] At some sites soil conditions would not support a septic tank leach field. In those cases the complex used an "oxidation pond," a large concrete-lined pit in which sewage was allowed to "oxidize." Crew members probably had to hope the wind was not blowing toward the air intake shaft.

▲ Sewage ejection pumps P6A and P6B. Sewage was pumped from this tank to a septic tank near the surface.

1 *Weapon System Familiarization,* Course OZR1821F/3121F, Technical Training, Shepard AFB, January 1965, 31

2 White, interview

3 In 1970 the Coordinated Universal Time system was devised by an international advisory group of technical experts within the International Telecommunications Union (ITU). The ITU felt it was best to designate a single abbreviation for use in all languages in order to minimize confusion. Since unanimous agreement could not be achieved on using either the English word order, CUT, or the French word order, TUC, the acronym UTC was chosen as a compromise. See http://tf.nist.gov/general/misc.htm

4 Department of the Air Force, *390 SMW Regulation* 55-7, paragraph 1-12

5 http://aa.usno.navy.mil/faq/docs/UT.html

6 The American Radio Relay League, *The Radio Amateur's Handbook,* 1966, 275

7 Operational Readiness Training Study Guide, 4315th Combat Crew Training Squadron, 1967, 4-5—4-9, TMM reference 20267

8 TO.21M-LGM25C-1, 1-106A

9 http://www.fas.org/nuke/guide/usa/c3i/saccs.htm

10 *Inter Complex Radio Communications,* a USAF document, authorship unknown, TMM archive reference 20084, 1969

11 TO.21M-LGM25C-1, 1-100-101

12 Glasstone, 520-522

13 TO.21M-LGM25C-1, 1-43

14 Ibid., 1-25, 27

15 Ibid., 1-140B

16 Ibid., 1-61, 66-67

17 Ibid., 1-56, 59-60

18 Air Training Command, *LGM-25 Maintenance,* Volume 3, 99-101

19 TO.21M-LGM25C-1, 1-72 Note: While the Dash One indicates seven tries were available some former crew members insist that one try was used in the process of installing the code. Six attempts is the number officially used by Titan Missile Museum docents.

20 Ibid., 67-68

21 Ibid., 1-72

22 A squib is a small explosive charge, in this case designed to puncture bladders full of electrolyte used to charge the APS batteries.

23 TO.21M-LGM25C-1, 1-133

24 Stumpf, 203-204.

25 Ibid., 63-67

26 Department of the Air Force, *390 SMW Regulation 55-7,* paragraph 1-5

27 *Weapon System Familiarization,* Course OZR1821F/3121F, Technical Training, Shepard AFB, January 1965, 40, 137

28 Morris, interview

29 TO.21M-LGM25C-1, 1-100, 101

30 Ibid., 1-75, 76

31 Batteries have a 5-hour rating at 210 ampere/hours. TMM archive reference number 32856

32 00-ALCP-144-4, A3-1s

33 *Weapon System Familiarization,* Course OZR1821F/3121F, Technical Training, Shepard AFB, January 1965, 135

▶ If the access portal collapsed during an attack, or it was unsafe to open blast door 8, the crew could open this escape hatch and climb a ladder to the surface. The hatch was normally closed and locked.

USAF PHOTO, TMM ARCHIVES.

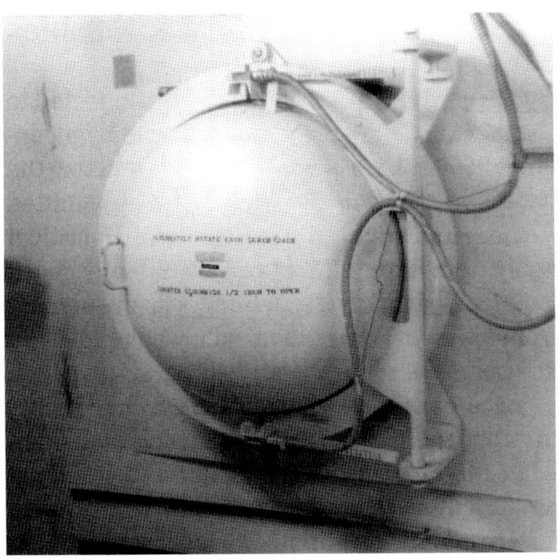

◄ A portion of the silo ladderway. This view looks up from level 4 to level 2.

USAF PHOTO, TMM ARCHIVES.

CHAPTER 6

THE SILO

The silo was composed of two elements, essentially two concentric cylinders: the inner cylinder, called the launch duct, with an inside diameter of 26.5 feet (8 m), which housed the missile; and the outer cylinder with an inside diameter of 55 feet (17 m). The space between these two cylinders was called the silo equipment area—a warren of pipes, cables and machinery needed to keep the missile in a constant state of readiness.

While the launch duct was hollow the silo equipment area surrounding it was divided into nine levels. Each level contained equipment for a specific purpose, and six levels had doors for access to the launch duct. A ladderway provided access to all levels of the silo equipment area while an elevator could access all levels except 1 and 9.

The structure of the silo was dictated by a number of engineering requirements that made it impossible to fully shock-isolate it in the same way the LCC was isolated from its surroundings. By necessity, the launch duct was connected to the silo at the base, and was braced by a concrete floor at level 3. Shock waves impacting the outer silo wall would, therefore, be transmitted to the launch duct. For that reason, all critical equipment in the silo was independently shock isolated—virtually everything was suspended from springs. Even the missile was cradled by an enormous set of shock absorbers.

Because silo shock isolation was impossible in any practical terms, engineers employed what might best be thought of as silo shock *management:* they designed the structures to absorb, damp and minimize shock transmission to the launch duct to the fullest extent possible. Engineering documents describe several methods of shock mitigation. These included designing the outer silo wall to permit it to flex above level 3; a slip coupling at the interface of the floor and the launch duct wall at level 3 to vertically uncouple the launch duct from the silo at that point; steel floors with gaps and slip joints so that shock energy would be transmitted to the launch duct only at level 3; and a gap between the top of the launch duct and the silo headworks.[1]

The outer wall of the silo was eight feet (2.4 m) thick from the surface to a depth of about 30 feet (9 m), at which point it tapered abruptly to a thickness of four feet (1.2 m). This taper permitted the silo to flex slightly, while the sheer mass of the eight-foot wall served to absorb lateral motion as well as vertical energy applied from above.

And it's all built for a single use. As one museum docent quipped, "It's a one-shot muzzle-loader." There was never any thought of reloading. The launch duct suffers surprising little damage during launch—nothing that couldn't be repaired in a few weeks. But the Wing had only one spare missile, and in the aftermath of a nuclear exchange, destruction of facilities and infrastructure would likely be so thorough and widespread there would be no means of installing another missile, even if you wanted to.

Launch Duct

The launch duct was a continuous uninterrupted shaft surrounding the missile. At the top was the silo closure door, and at the bottom were two exhaust ducts and the flame deflector plate, usually referred to as the "W" because of its shape. At six levels between the top and bottom (levels 1 through 5, and 7) retractable waffle-like work platforms could be lowered to give maintenance crews direct access to the missile. Levels 1 through 4 and level 7 each had eight platforms while level 5 had only

▼ This aerial photo provides a splendid view of the launch duct and exhaust shafts. Note cooling towers and hardstands.

TMM ARCHIVES.

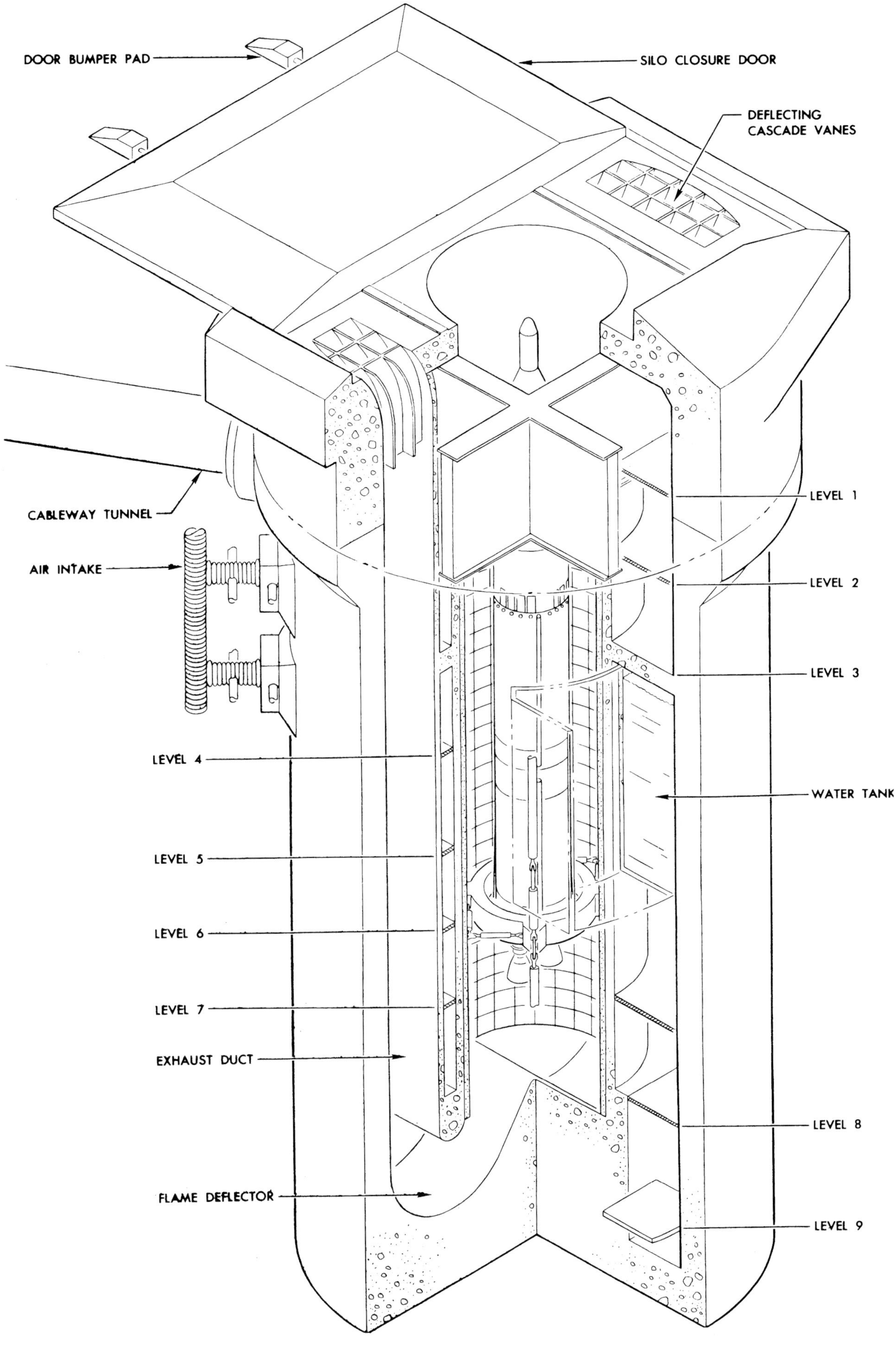

Two views of the silo and launch duct.

◀ At left, a detail from a Martin Company illustration.

▶ At right, a drawing from the Dash One. The Dash One image is rotated about 90 degrees from the Martin drawing and provides an additional perspective of the flame deflector and exhaust shafts. Curiously, the newer Dash One image shows the missile fitted with the older Mark IV reentry vehicle, which was never deployed on Titan II.

TMM ARCHIVES.

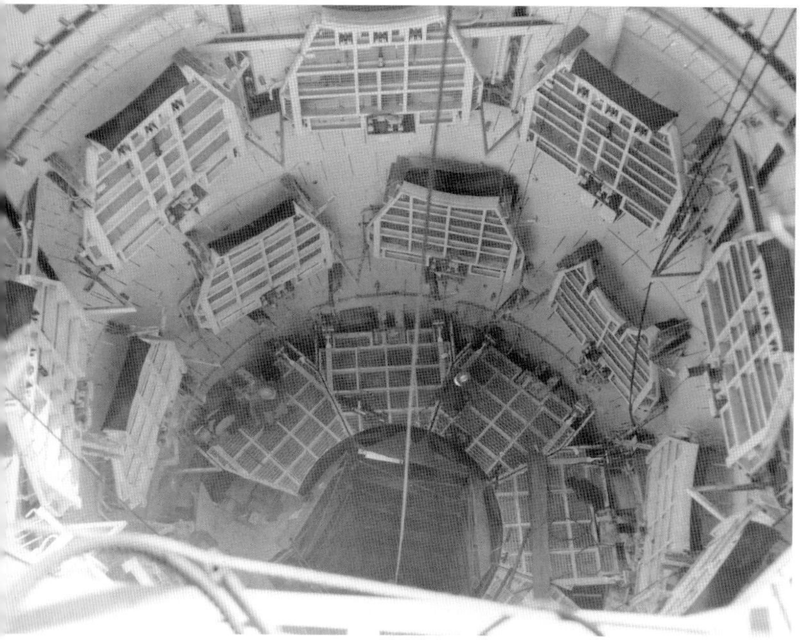

▲ Work platforms as seen before installation of the missile. In this image the platforms on level 5 have been lowered.

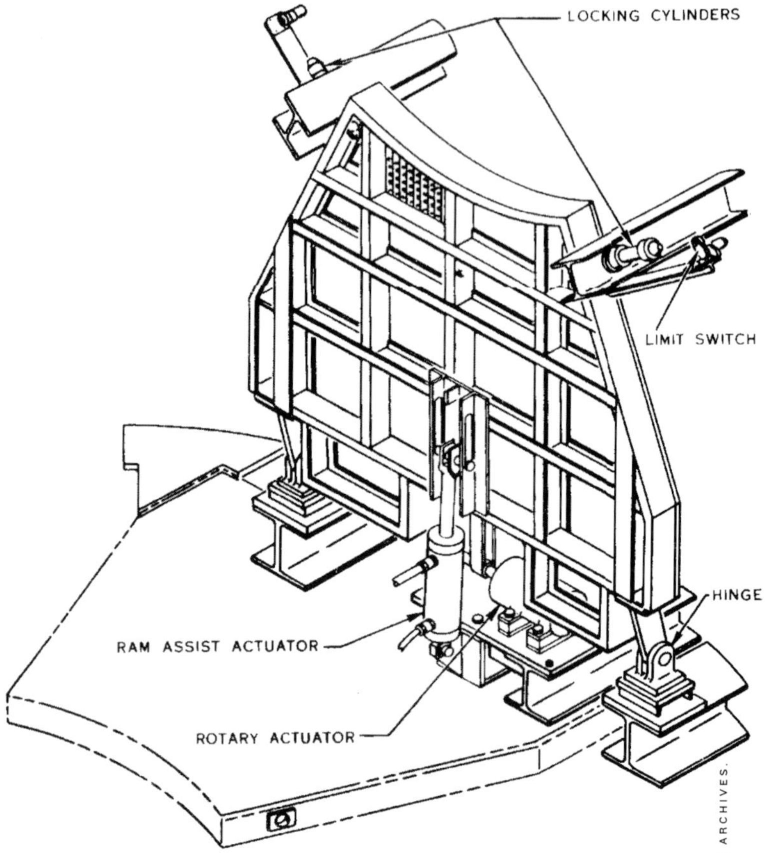

TMM ARCHIVES.

▲ Schematic showing platform, hydraulic motor and locking pins.

four. The work platforms were normally kept in the up position. The Titan Missile Museum has lowered two of the platforms on level 2 to demonstrate how maintenance teams might have worked near the missile. In reality, whenever workers needed to enter the launch duct all eight platforms would be lowered, creating a floor completely surrounding the missile. With that in mind it is important to note that if even one of the work platforms was down the missile could not be launched. An inability to launch, for whatever reason, resulted in the complex being off alert, and being off alert was not a good thing.

The wall separating the launch duct from the silo equipment area was 18 inches (46 cm) thick and was steel plated on both the inner and outer surfaces. Additionally, the interior surface of the launch duct was covered with special acoustic modules. These large rectangular boxes were about 12 inches (30 cm) thick and made of aluminum and fiberglass and were intended to muffle the intense roar of the engine during launch. The concern was that vibrations caused by the noise would damage the missile before it cleared the silo. Because the acoustic modules alone were thought to be insufficient protection, engineers came up with an additional measure to make certain the missile would not be damaged.

Near the bottom of the launch duct just above the flame deflector were four huge water nozzles for the water deluge system. During a launch, about 30 seconds before engine ignition, the water deluge system began flooding the flame deflector at the rate of about 150 gallons (560 l) per second.[2] When the engine ignited, this water vaporized, absorbing a tremendous amount of acoustical energy in the process. The exhaust of Titan II's engine was transparent—

USAF PHOTO, TMM ARCHIVES.

▶ Construction photo shows acoustic modules lining the launch duct wall. The aluminum-framed modules were stuffed with fiberglass.

PHOTO BY JOHN DESMOND.

◀ Two of the four water deluge nozzles used for additional sound suppression. They were gravity-fed from a large in-silo water tank. Note level 7 work platforms at top of image.

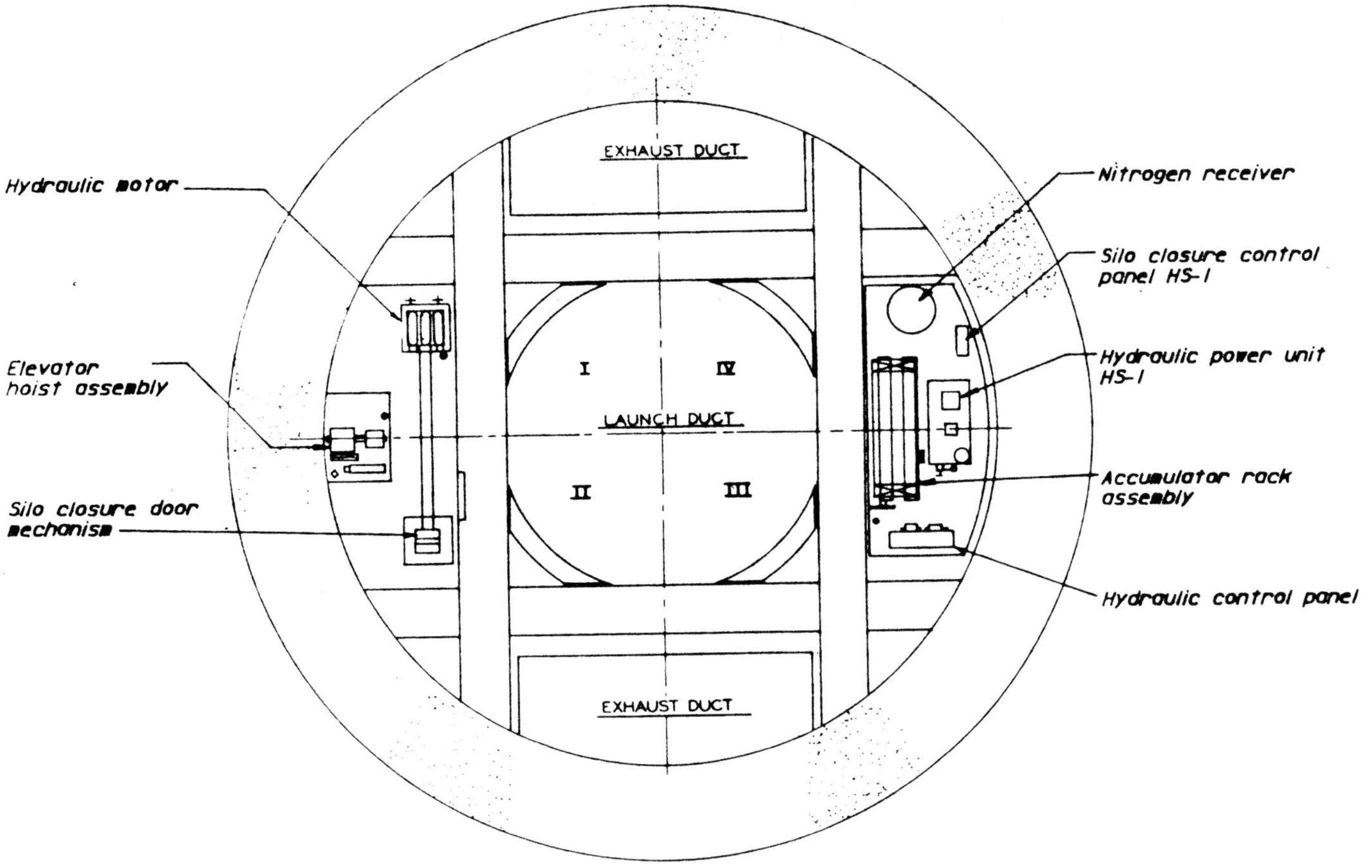

SILO EQUIPMENT AREA - LEVEL I

TMM ARCHIVES

virtually all the "smoke" seen billowing from the exhaust ducts during a launch was actually steam from the vaporized water of the deluge system.

Silo Equipment Level I

Level 1 was located at the top of the silo just beneath the closure door and was subdivided into three smaller parts: Levels 1A, 1C and 1D (level 1B was the area inside the launch duct at the same level). Level 1 was not accessible from the elevator, though the elevator could be parked at level 1A for servicing by maintenance crews. Ladders from level 2 led up to levels 1A and 1C, with an additional ladder from 1A to 1D.

▶ Twin five-cylinder hydraulic motors opened and closed the silo closure door. A drive shaft extends to upper left.

Level 1A housed the hydraulic motors used to open the silo closure door. The motors developed a total of only 40 horsepower but were geared down to provide very high torque. A large drive shaft extended across the length of level 1A. On each end of the shaft was a pulley that wound and unwound the large metal cables that actually pulled the door open and closed. By necessity this level was not shock isolated. If it were, the much heavier door would tend to pull the much lighter motor upward on the springs instead of the motor pulling the door open. Lack of shock isolation here was one of the few weak spots in Titan II's armor.

Just behind the closure door drive motor, a short ladder led to level 1D. This level contained the control system and motor for the silo elevator. Finally, a small door on level 1A provided access to the missile at the level of the reentry vehicle (RV).

Level 1C was on the opposite side of the silo from level 1A. It housed hydraulic system one (HS-1) that provided power for the silo closure door motor.

At the core of HS-1 was a group of cylinders called accumulators. One end of each accu-

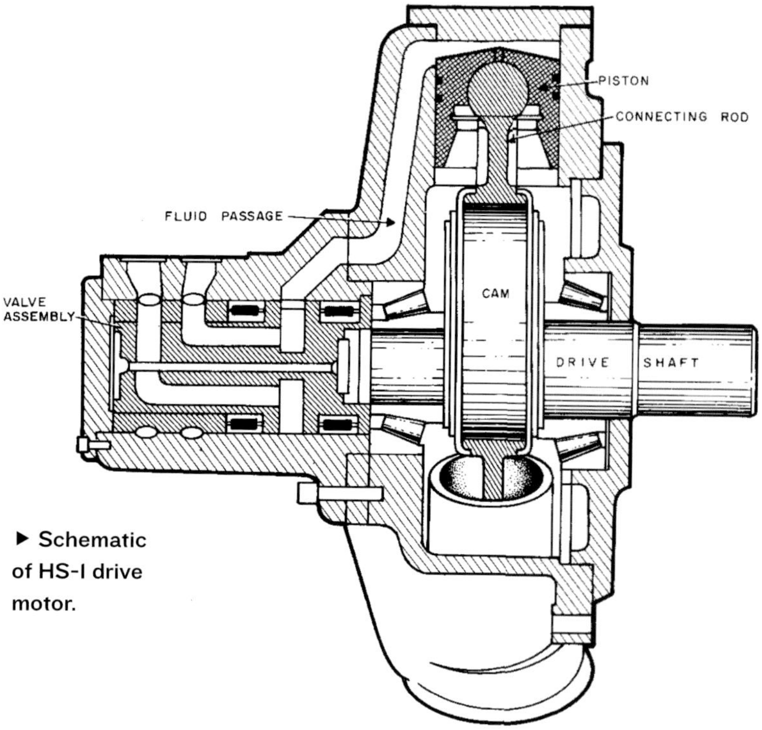

▶ Schematic of HS-I drive motor.

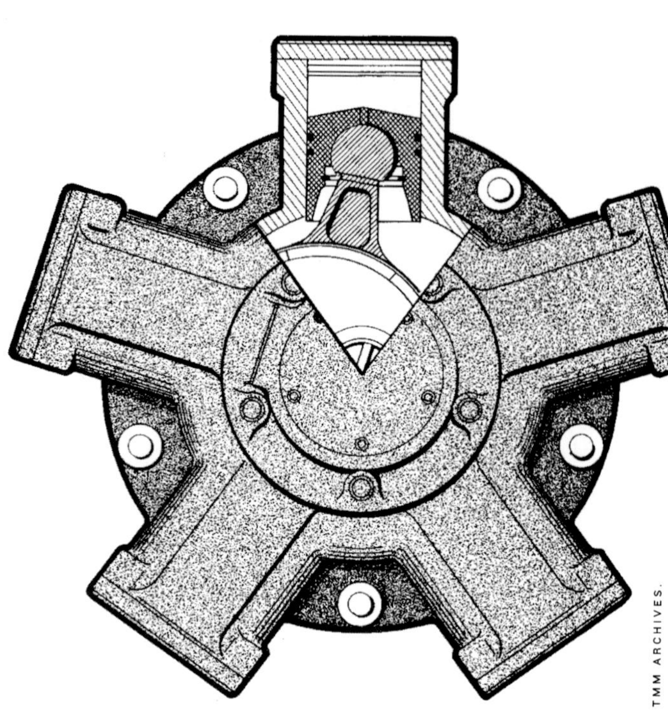

TMM ARCHIVES.

mulator was connected to a tank containing high-pressure nitrogen gas. The other end was connected to a hydraulic pump. Inside the accumulators a piston separated the gas and hydraulic fluid. Pumping hydraulic fluid into the accumulators pushed the pistons to one end, compressing the nitrogen gas on the other side. This compression essentially cocked the trigger. Extending the analogy, one pulled the trigger by opening a valve on the hydraulic side. This relieved the pressure on the pistons permitting the nitrogen gas to flow back into the accumulators, pushing the pistons and forcing hydraulic fluid into the drive motor.[3,4]

In a nuclear attack one of the very first casualties would be the commercial electric power grid. Electric power would fail nationwide within the first few minutes of a nuclear attack because of the effects of EMP. HS-1 enabled the door to be opened and closed without relying on vulnerable commercial or back-up generator power. The only electricity required was that needed to operate the valves, and that could be supplied with battery power alone. The accumulators stored enough energy to operate the door though one and a half cycles (open-close-open) without having to run the hydraulic pump electric motor.[5]

Silo Equipment Level 2

The silo equipment area and launch duct were ventilated and air conditioned with a bewildering assortment of pumps, fans, chillers, blast valves, dampers, filters, heaters, coolers, air washers, dehumidifiers and heat exchangers.

Components connected to this complex tangle of pipes and ducts were located on

▼ Located on a half-level just above the HS-I drive motor was the motor and control unit for the silo elevator.

▼ Hydraulic power supply unit for the HS-I drive motor. Manual control panel (left), HS-I hydraulic pump and motor (second from left), electric valve panel (center) and accumulator rack (right). The entire level is suspended from springs.

◀ HS-I accumulator rack (left) and nitrogen receiver tank (center). Note shock isolation canisters flanking the receiver tank.

almost every level of the silo equipment area, and the first glimpse of the system could be found on level 2. Three large shock-isolated water chillers provided cold water for several air conditioners. Each unit had a chilling capacity of 25 tons (300,000 BTU). Two of the chillers were operational, while the third was for standby use. The chillers provided cold water to several air conditioners needed to maintain the complex and missile at the correct temperature.

The chillers operated in conjunction with cooling towers on the surface until the late 70s when they were made unnecessary by the installation of modern chillers located topside. After that, the Trane chillers were relegated for use only under plumbing shutoff conditions employed during advanced stages of readiness.

In plumbing shutoff mode, all pipes leading

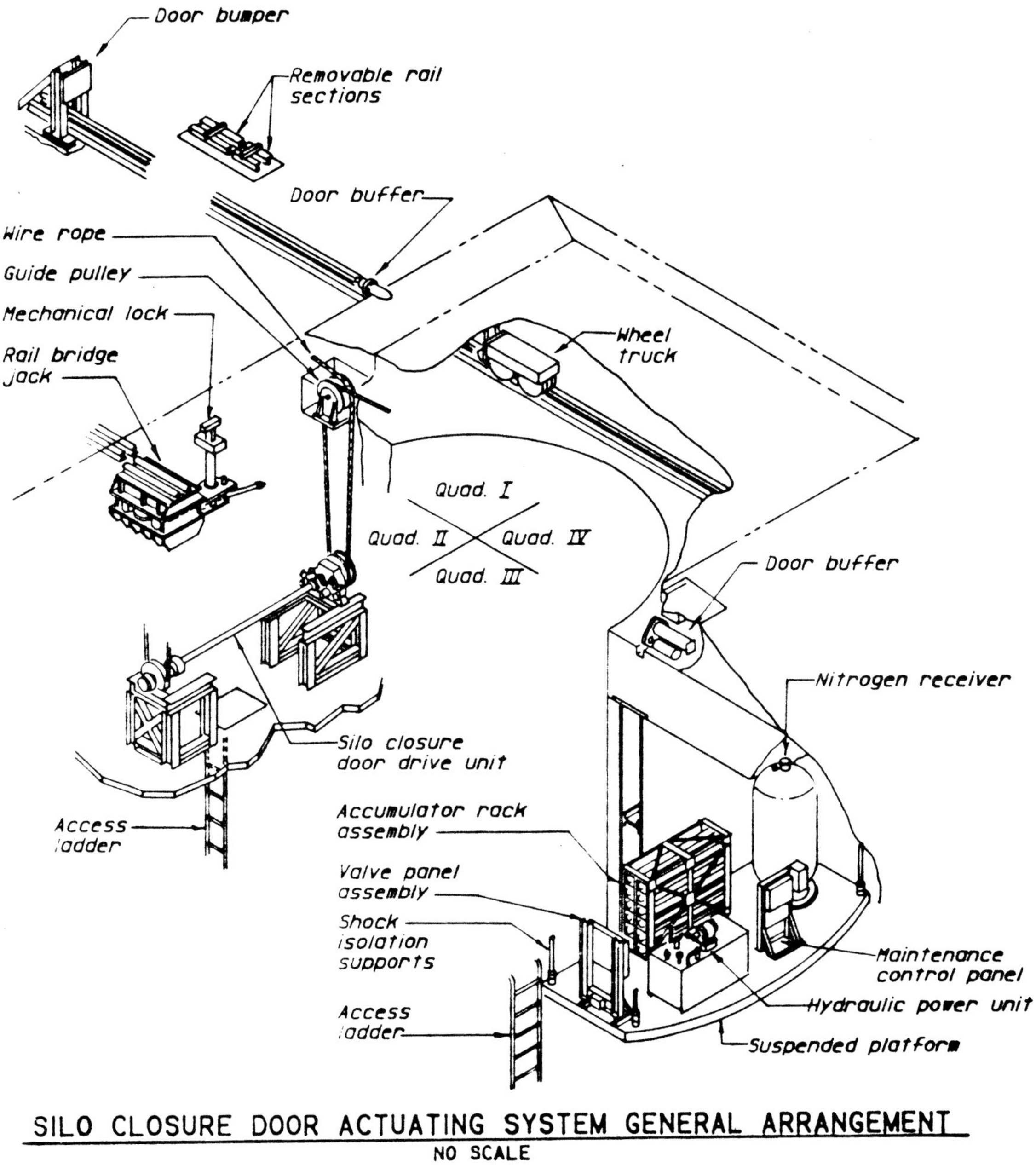

TMM ARCHIVES.

▶ Diagram of HS-I and basic configuration of door motion components.

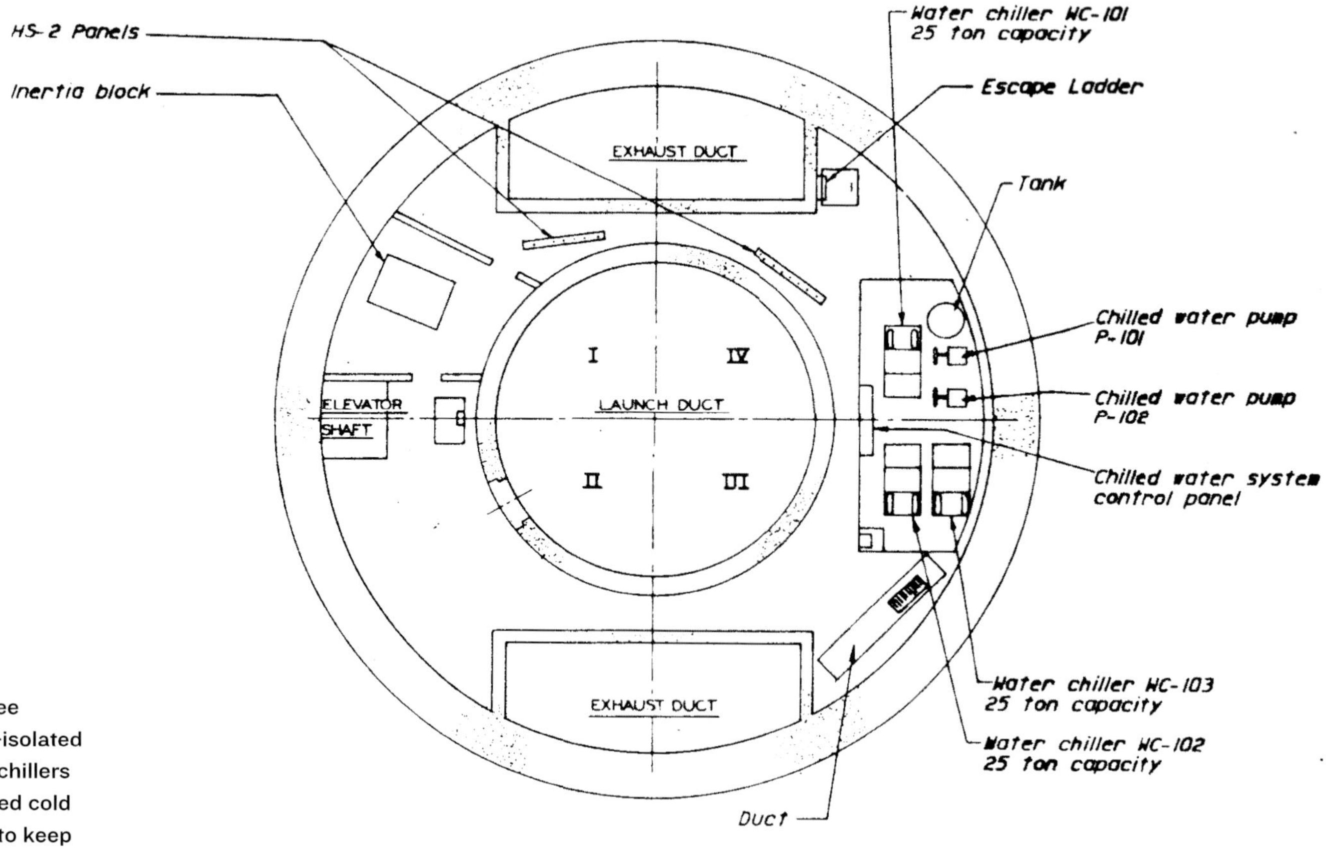

SILO EQUIPMENT AREA – LEVEL 2

TMM ARCHIVES

▼ Three shock-isolated water chillers provided cold water to keep both propellant and crew comfortable. The units were made by Trane and had a rated chilling capacity of 25 tons each. Two were operational, one was standby.

into or out of the underground portion of the complex would be shut off with pneumatically actuated valves called plug valves. These valves prevented shockwaves from entering the complex though plumbing lines. There were 11 such plug valves around the complex.[6] In a plumbing shutoff scenario, heat normally dissipated by the topside chillers would instead be dissipated into a 100,000-gallon (380,000 l) water tank in the silo. This mode of operation could be used only until the water in the tank became too warm to be an effective heat sink—about 24 hours.[7]

Prior to the installation of a more advanced guidance system in 1978, silo level 2 contained a special piece of equipment called the azimuth

◀ Typical plug valve. All plumbing lines could be shut off to prevent shock waves from entering the complex via piping to or from the outside world.

USAF PHOTO, TMM ARCHIVES.

▲ A rare photo of the azimuth alignment set (AAS) auto-collimator unit. It looked much like an old-fashioned slide projector. Its light beam projected into a window in the missile's IMU and kept the gyroscopes pointed north. The auto-collimator was not needed by the new guidance system and was removed in 1978.

alignment set (AAS) auto-collimator. The AAS provided a constant azimuth reference for the missile's inertial measurement unit (IMU), the device that told the guidance computer how to steer the missile. In effect, the AAS constantly told the IMU the direction of north. An accurate bearing on north was critical for the missile in finding its way to the target.

The auto-collimator was mounted on a large concrete platform. This platform was actually formed as part of the silo wall, and protruded from the silo wall on level 3. A beam of light from the AAS was directed through a porthole in the silo, through an open hatch on the side of the missile and into a window in the IMU.

Inside the IMU a prism reflected the beam back to the auto-collimator where a detector compared the out-going beam to the reflected beam. If the two beams did not align exactly, it meant the gyroscopes in the IMU were drifting. The AAS then sent a signal to the MGACG in the LCC, which in turn sent a signal to the IMU saying, essentially, "tweak the gyros."

The auto-collimator itself required periodic calibration. This was accomplished with a set of precisely located calibration monuments located topside and a theodolite (a type of surveying instrument). The theodolite was used to get a bearing on the monuments, which were located several hundred yards (meters) outside the security fence. The monuments were about five feet (1.5 m) tall and 18 inches (45 cm) in diameter, and a metal plate on top could be fitted with a sighting target. The theodolite was then turned a specific number of degrees and sighted down a slanted pipe to the auto-collimator, which was carefully aligned to the theodolite's crosshairs.[8] Calibration was required every six months or as dictated by technical orders. Unlike with Atlas and Titan I, the north star was not used to calibrate the AAS.

▼ Diagram showing arrangement of AAS and IMU.

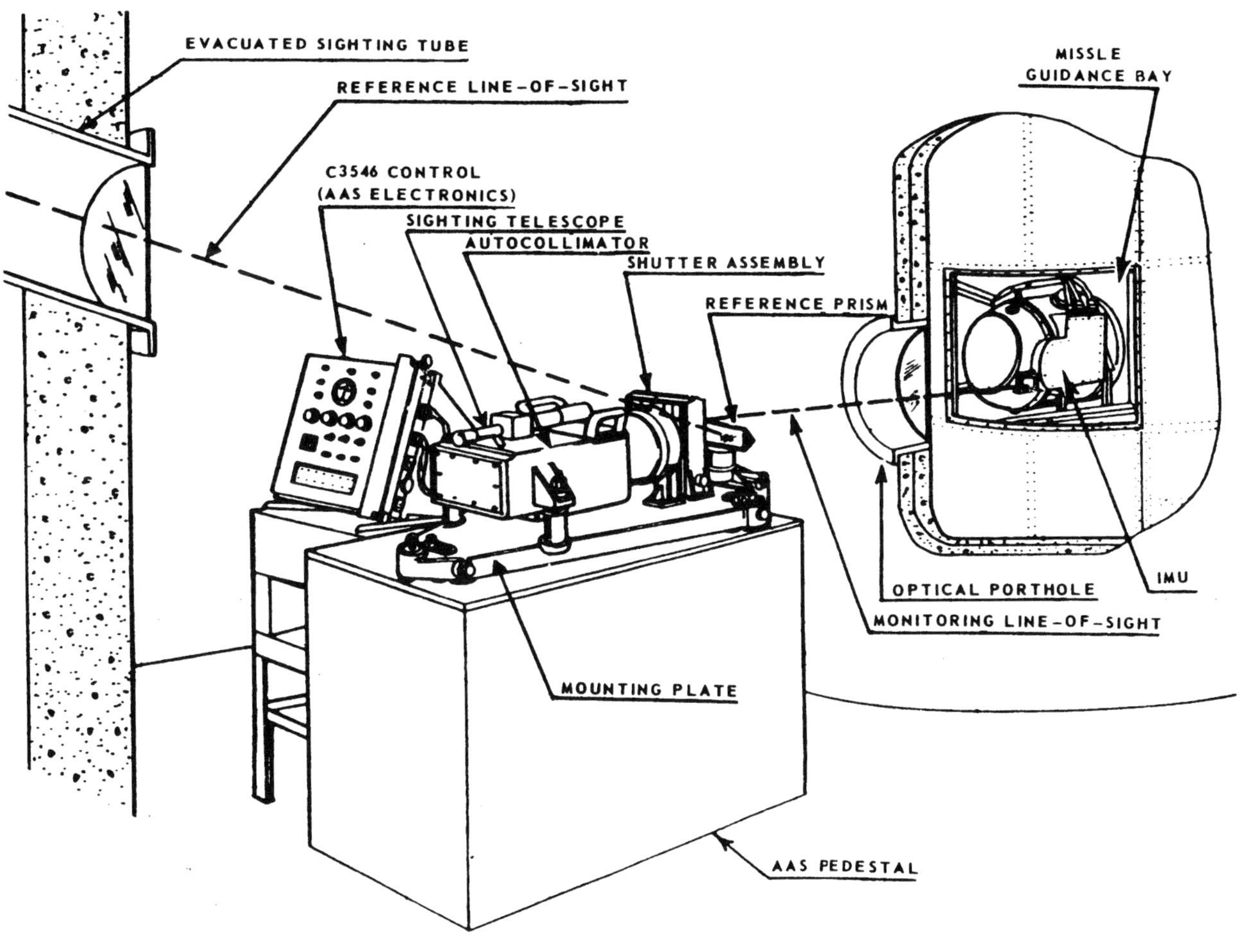

Figure 2. Launch complex collimator room AAS layout.

TMM ARCHIVES.

▶ Technique for calibrating the AAS. Calibration was required every six months or as dictated by technical orders.

1. Electronic equipment shelter
2. Transfer theodolite
3. Transfer theodolite tripod
4. Survey target (two places)
5. Survey target station (two places)
6. MX-3880 optical cell assembly
7. Sight tube
8. MX-3879 optical cell assembly
9. MX-3384 autocollimator
10. MX-3388 shutter
11. MX-3385 prism
12. MX-3881 optical cell assembly
13. MX-2662 mounting plate
14. Vertical leveling mirror
15. MT-2643 mounting plate
16. Autocollimator mount
17. Coordinate autocollimator
18. C-3546 control
19. Vacuum pump and motor assembly
20. Survey transfer station

▲ One of two AAS calibration monuments. Author for scale.

Both the computer and IMU of the old guidance system required constant cooling. This was provided by an air conditioner suspended from the ceiling and located directly in front of the entrance of the cableway. This unit was connected to the missile by a large flexible hose.

The AAS auto-collimator was located in what was then a separate room on level 2. An air conditioner suspended from the ceiling kept the room between 70-90 °F (21-32 °C) as required to maintain a dew point of 50 °F (10 °C) or less. This prevented condensation on the AAS lenses.[9]

The need for the AAS vanished with the installation of the new guidance system, and the auto-collimator was removed along with its air conditioner and the air conditioner for the computer and IMU. There was no need to remove the concrete platform and topside monuments and they were abandoned in place.

Silo Equipment Level 3

The emergency generator and main electric power distribution equipment were located on level 3. The generator was driven by a six-cylinder, 510-horsepower diesel engine, and could supply 350 kilowatts of power.[10] Its huge cylinders had a bore of 8½ inches (22 cm) and its pistons a stroke of 10½ inches (27 cm).[11] While even the Dash One refers to the generator as being supercharged, technically, it was turbocharged.[12] The unit consumed about 25 gallons (95 l) of fuel per hour under an average load.[13] No glow-plugs were used. To ensure an easy start, an electric immersion heater circulated hot water through a jacket around the generator's engine.[14] For added reliability the engine was started with compressed air supplied from a compressor on level 7.[15]

While 350 kilowatts may seem like plenty of electricity, it was far less that the 500 kilowatts

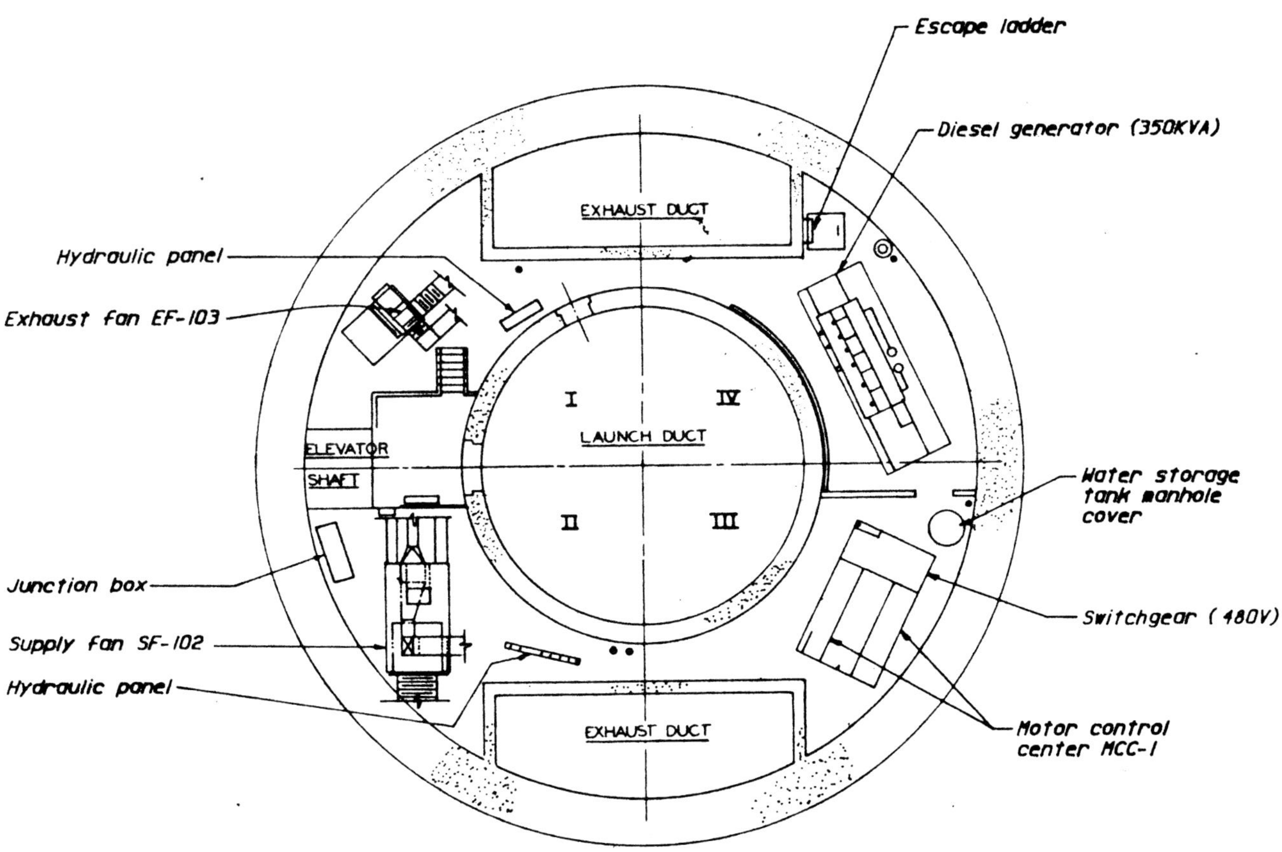

SILO EQUIPMENT AREA–LEVEL 3

TMM ARCHIVES.

◄ The 350-kilowatt diesel generator.

normally available to the complex from commercial power, and certain conservation measures were required when running on generator power.[16] For example, some normally automatic systems would need to be started manually. These included hydraulic systems and a large fire fighting pump. In the early years of the program the generator was tested every Sunday morning. Later on, the generator was tested only as directed by higher headquarters or maintenance requirements The entire generator unit, including its shock isolation platform, weighed 14 tons (12.7 metric tons).[18]

Just as in an automobile engine, the generator's engine was cooled by water. But unlike an automobile, the generator had no radiator to dissipate the heat it created. Instead, as the engine's heat boiled the water, additional water was pumped in. Steam from this cooling system was fed into a centrifugal steam separator that separated the steam from any remaining liquid water. Liquid water was reused while the steam was mixed with the exhaust and, under normal conditions, was blown out the silo exhaust airshaft. A stream of water was injected directly into the 700° F (370° C) exhaust to prevent the exhaust shaft blast valve from overheating.[19] The engine used about 1.5 gallons (5.5 l) of cooling water per minute.

In the event of nuclear detonation, the silo blast valves would close automatically due to blast overpressure. If the blast valves closed while the generator was running, the steam and diesel exhaust would be vented directly into the silo equipment area.[21] The blast valves were set to automatically reopen 90 seconds later. Should

▲ Diesel exhaust was routed through BV-I and blown to the surface. An exhaust water injection system helped cool the valve.

► Upon loss of commercial power the switchgear unit automatically started the generator and transferred power.

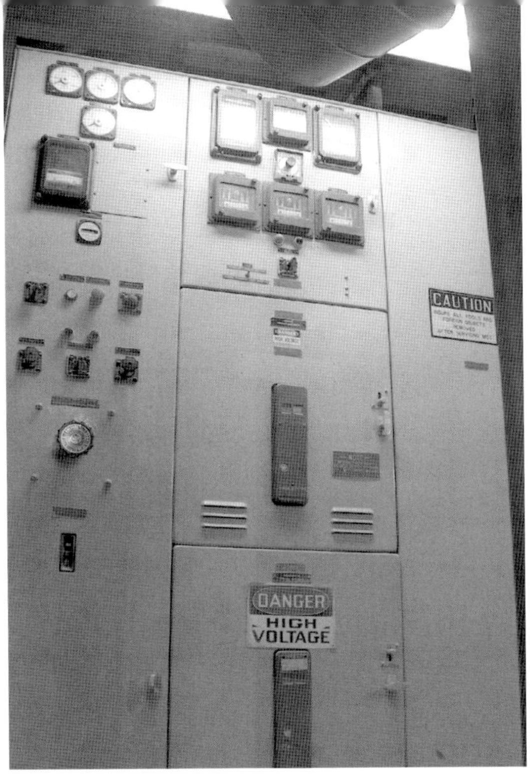

▲ A portion of MCC-I, the main power distribution center for the entire complex.

the valves fail to reopen, however, it is not clear how long the generator could be operated, as it would eventually deplete the oxygen in the silo, causing the engine to stall. And even in just 90 seconds, venting exhaust into the silo equipment area would have the effect of making entry into the silo impossible without breathing apparatus.

Near the generator was a large electrical cabinet containing what was called the switchgear, and a large walk-in electrical distribution panel, designated motor control center one (MCC-1). The switchgear equipment was used to start the generator and transfer the complex to backup power. These functions could be accomplished automatically by the switchgear or manually from the FPCB in the LCC. In the automatic mode, a 10 second delay prevented the generator from starting in response to momentary outages. Transferring from the generator back to commercial power had to be done manually and could be done from controls at the switchgear or the FPCB.

MCC-1 contained a number of large lever switches that distributed power to the entire complex, including power to MCC-2 on level 3 of the LCC. If commercial power was lost and the generator failed to start, the result was referred to as a "complete FAC" (complete failure of AC power). A complete FAC rendered the complex dark, except for a few battery operated emergency lights, and completely silent—an eerie and unnerving state of affairs. One crew person who experienced such an event reported the only sound heard was the murmur of water draining down the many pipes in the complex. Another reported that while on his sleep shift he was once abruptly awakened by the silence.

Tucked away behind the generator was a manhole for access to a tank that held 100,000 gallons (380,000 l) of water. This gigantic tank wrapped around the east side of the silo and extended from the floor of level 3 to the ceiling of level 7. It supplied the sound suppression water deluge system used immediately prior to a launch, and it could be used for launch duct fire suppression. It was also used by the chillers, as cooling water make-up (for losses due to evaporation), for the diesel cooling and exhaust water injection system, and as a heat sink for the water chillers during plumbing shutoff

◄ Access hatch to the silo water tank. Divers inspecting the tank had to be lowered in with a boson's chair or chain ladder. Note "hot stick" for rescuing someone who is being electrocuted. The generator and MCC-I are just out of view (at left and right, respectively).

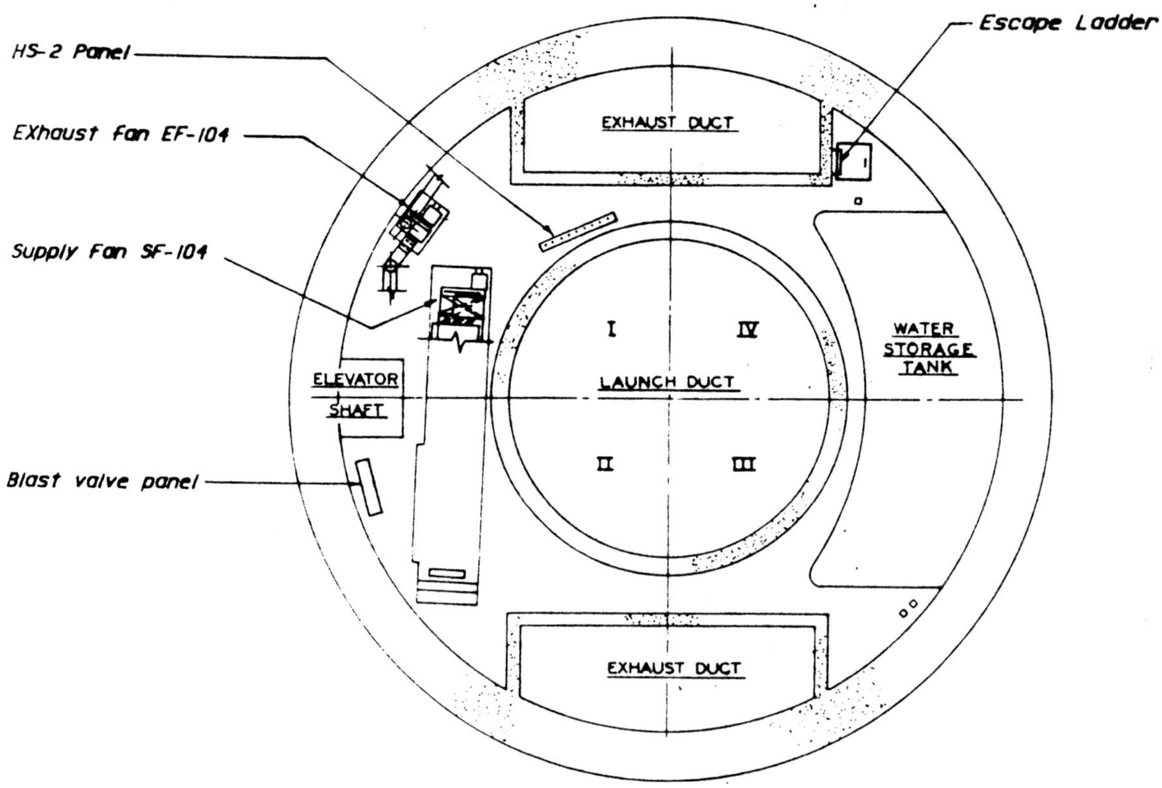

SILO EQUIPMENT AREA-LEVEL 4

TMM ARCHIVES

conditions.[22] Finally, this water could also have been used to wash down propellant spills or to generate foam for a fire suppression system on level 5. The water in the 100,000-gallon tank was never used as a source of domestic water for the crew. Water for drinking or cooking came directly from the soft water storage reservoir on the surface and was stored in the pressure tank on LCC level 1.

The 100,000-gallon tank is often referred to as the "hard water tank" to distinguish it from its soft companion on the surface. While the tank enjoyed some protection by virtue of being underground, it was not, strictly speaking, hardened. Because it was attached directly to the silo wall and was not shock isolated, the tank was subject to the effects of shock waves impinging the silo. In a significant shock event the tank may well have ruptured, or at minimum, sprung significant leaks. This would have been a problem only if the missile had not yet been launched and was still thought to be launchable. If the missile had been launched prior to tank rupture, or if the missile had sustained damage, a ruptured tank would have been a moot point, except that it would have made use of the generator problematic, depending on the extent of the leakage.

Silo Equipment Level 4

Directly in front of the silo elevator on level 4 was an air washer (AW-101), essentially an evaporative cooler, which was originally used as the silo equipment area cooling system. Almost immediately it began to wreak havoc with the silo humidity level, and was quickly converted to a conventional air conditioning system with the installation of cooling and heating coils to replace the original water spray.

Silo Equipment Level 5

An air conditioning system on level 5 (AC-102), which derived it "cold" from the chilled water system, cooled the launch duct and missile. AC-102 worked in concert with a fan (EF-102), also on level 5. While designated as an exhaust fan, EF-102 normally served

▼ AW-101 was almost immediately converted to a conventional air conditioner when the excessively high humidity levels it created led to major corrosion problems in the silo.

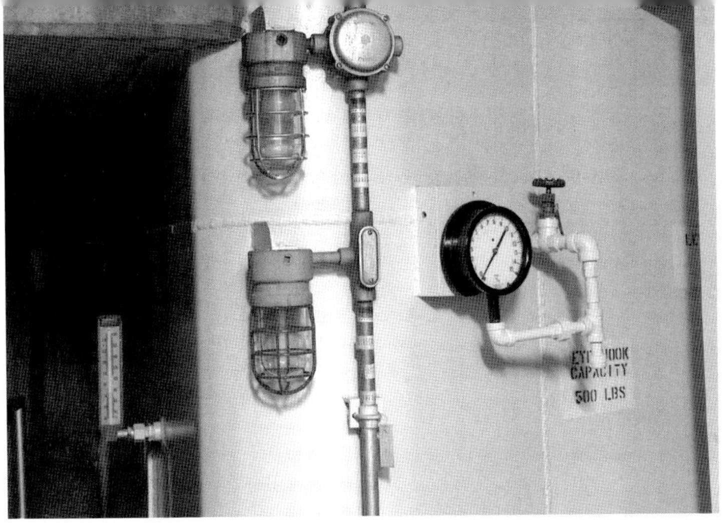

▲ A simple pressure gauge and thermometer displayed the level and temperature of water in the silo tank.

▲ An automatic level control kept the silo water tank topped off. On rare occasions it would malfunction, causing the tank to overflow.

to recirculate air from the launch duct back through AC-102.[23] Maintaining the launch duct and missile at the proper temperature was among the most critical tasks on the complex, so it is curious that AC-102 had no backup system. Of course, there wasn't much to go wrong. The most likely mode of failure would have been a malfunction of the motor, or of the bearings for the fan or chilled water pump. While maintenance crews were always on call, it is not difficult to imagine a series of "Murphy-like" events that might have seriously delayed repair.

Level 5 also housed a 2500-gallon (9500 l) diesel fuel tank used to refill the generator's 55-gallon (200 l) "day tank" on level 3. This 2500-gallon tank could in turn be refilled from an 8000-gallon (30,000 l) tank near the surface.

Filling the tank on level 5 was a simple

◄ A bank of mercury lamp ballasts for the launch duct lighting system.

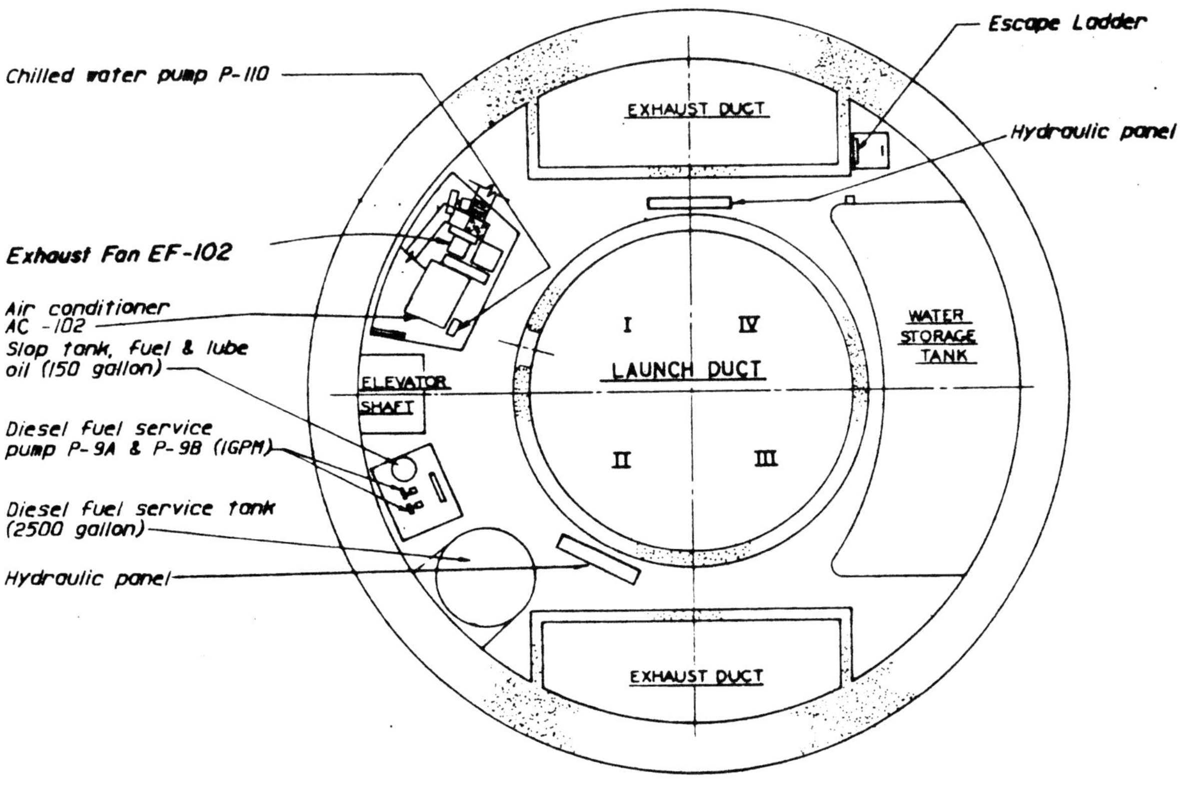

SILO EQUIPMENT AREA–LEVEL 5

TMM ARCHIVES

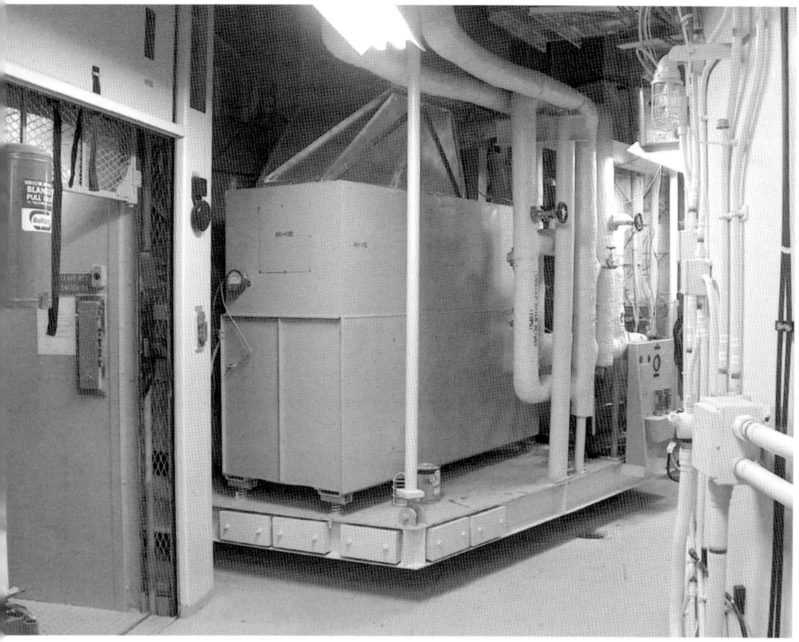

▲ AC-102 kept the launch duct and missile at the crucial 60 °F (15.5 °C) mark. This critical unit had no backup. Note silo elevator at left.

◀ Diesel fuel storage tank transfer pump timer located on the first landing of the access portal stairway. Another mystery.

matter. A switch adjacent to the tank activated a timer that turned on the fuel transfer pump for 15 minutes. This was enough time to pump approximately 75 gallons (280 l) into the tank.[24] The timer system was added later to reduce the chance of over-filling, although there was an overflow "slop tank" just in case.

What is somewhat perplexing about this scheme is the location of the timer. It was located on the first landing of the access portal. This meant that if the portal collapsed during an attack, the timer would be damaged, and the crew would be unable to refill the underground tank. Even assuming the portal did not collapse, the timer would almost certainly be disabled from the effects of EMP. Equally perplexing is the fact that the underground fuel tank was not shock-isolated—it was attached directly to the silo wall. This curious detail is made more curious still by the fact that even the fuel tank's overflow tank was shock-isolated, as were the pumps that supplied the fuel to the shock-isolated generator.

The lack of shock isolation for the 2500-gallon tank, combined with the soft nature of the timer and topside tank made post-attack use of the generator problematic.

One might argue that any shock strong enough to damage the underground diesel tank would also disable the missile and take the site out of action, and so what difference did it make whether the tank was shock isolated or not?

▶ Diesel fuel storage tank, with overflow tank at far right. The overflow tank and fuel pumps were shock isolated. The main tank was not. Puzzling.

It would make a huge difference to the crew. Setting the mission aside for a moment, if there was a nuclear shockwave strong enough to disable the missile, the crew would certainly want to stay underground for a couple of weeks at minimum, waiting for the radioactive fallout to subside to less-than-lethal levels before venturing topside for a look around. In this case they would want to use the generator to run the lights now and then so as not to spend the entire time in total darkness (the battery operated emergency lights would last only a few hours). Moreover, the crew would certainly want to try the radios in an effort to contact the outside world. The generator would be essential for that. And the generator would be essential in running the air supply fan if the crew chose to risk running it to pump fresh air into the control center.

Here are a few important points regarding the generator when commercial power is not likely to be available again soon, or ever:

- The 8000-gallon topside tank was unlikely to be available after an attack, limiting the crew's fuel supply to 2500 gallons.
- Without the 2500-gallon tank, the generator had only enough fuel for a couple of hours of operation.
- When not operating, the generator was kept warm with an electric heater. Without commercial power, the heater could not work. The longer the generator went without its internal heater, the more difficult it became to start.
- Starting the generator depended on an air compressor, which in turn depended on the generator—a disquieting circularity. If the compressor's air tanks were exhausted in an attempt to start the generator, there was no chance of starting it.

Shock isolating the diesel tank, relocating the timer, and providing even a modicum of hardening for the topside fuel tank and associated pump would not have posed any significant engineering challenges nor would it have added significantly to the cost of construction, so it is difficult to understand why engineers chose not to do it. From the perspective of crew survival in a post-attack scenario, it seems like a fatal flaw in design.

Silo Equipment Level 6

Level 6 supported the equipment used to monitor for fuel and oxidizer vapors. This equipment was made by the Mine Safety Appliance

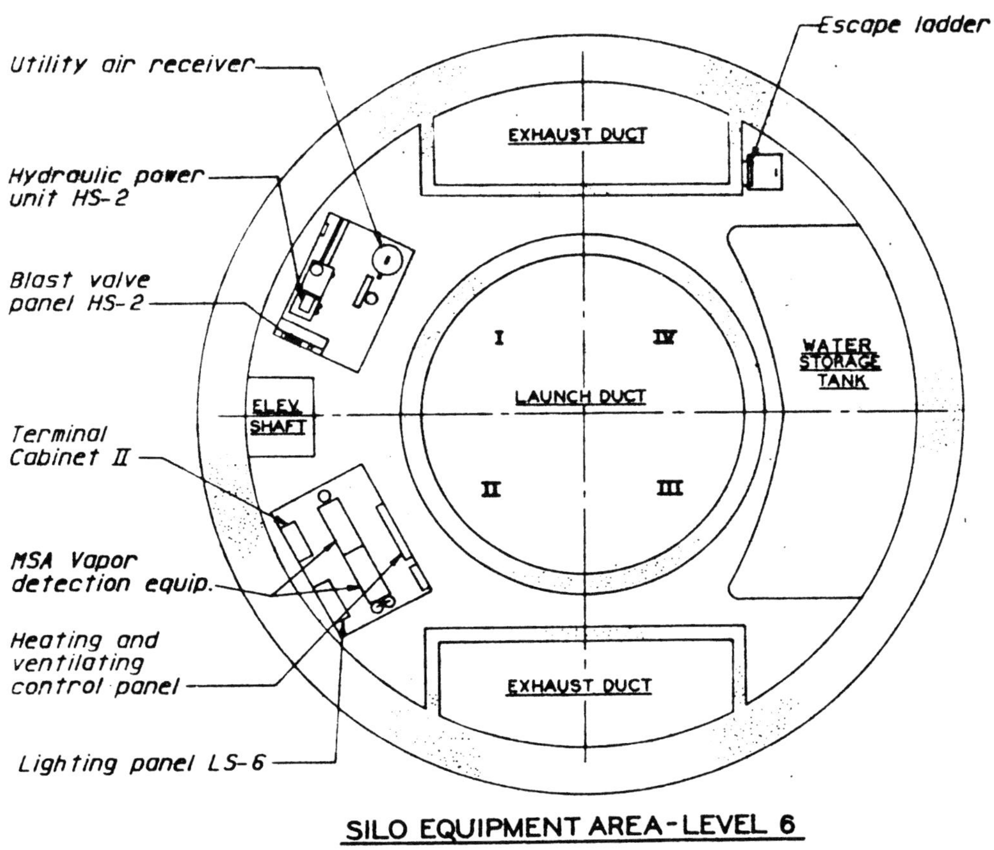

◄ The MSA was a self-contained automated chemical lab. It constantly analyzed air samples from around the silo and launch duct, checking for telltale propellant vapors. The large plastic bag was used to calibrate the unit. A manual-override panel for the silo ventilation system is at left.

Company and was universally referred to as the MSA. The MSA worked with a gaseous reagent chemical and a radioactive ionization chamber to perform a continuous real-time chemical analysis of the air in specific areas of the silo and launch duct. Data from the MSA was sent to the VDAP display panel in the blast lock and could trigger alarms in the LCC.

Near the MSA was a manual override control panel for the silo and launch duct ventilation equipment.

Hydraulic system two (HS-2), also located on level 6, was used to operate the launch duct work platforms and silo blast valves.

◄ Backside view of the MSA. Manual-override sampling controls are at top. Large meters replaced the original chart recorders for display of vapor levels. Normal readings were very close to zero.

▲ HS-2 powered the silo blast valves and work platforms. Note fire suppression foam tank on the wall at left.

▲ Under normal conditions the domestic water pumps pressurized both the domestic and industrial water systems. One unit was operational, the other was standby.

Behind HS-2 was the foam tank. This tank contained ox blood (yes, real ox blood) that could be mixed with water to create foam for fire suppression in the diesel fuel storage tank area on level 5. The foaming process was automatic and was initiated by fire detectors near the storage tank. Fire detectors were strategically located throughout the complex. Interestingly, modern smoke detectors first became available in 1969 yet were never incorporated into the site's fire protection system.

Silo Equipment Level 7

There were two separate and distinct water systems in the complex: the domestic water system, which provided potable water for the crew, and the industrial water system, which provided water for all other purposes. Under normal conditions, the 7.5 horsepower domestic water pumps, P-3A and P-3B, pressurized both systems. Although the industrial and domestic water systems were supplied from the same pumps, they were isolated from each other by a backflow-preventing check-valve.

Pumps P-3A and B were not shock isolated since in an advanced state of readiness the complex would be operating in plumbing shut-off mode, during which the water line supplying

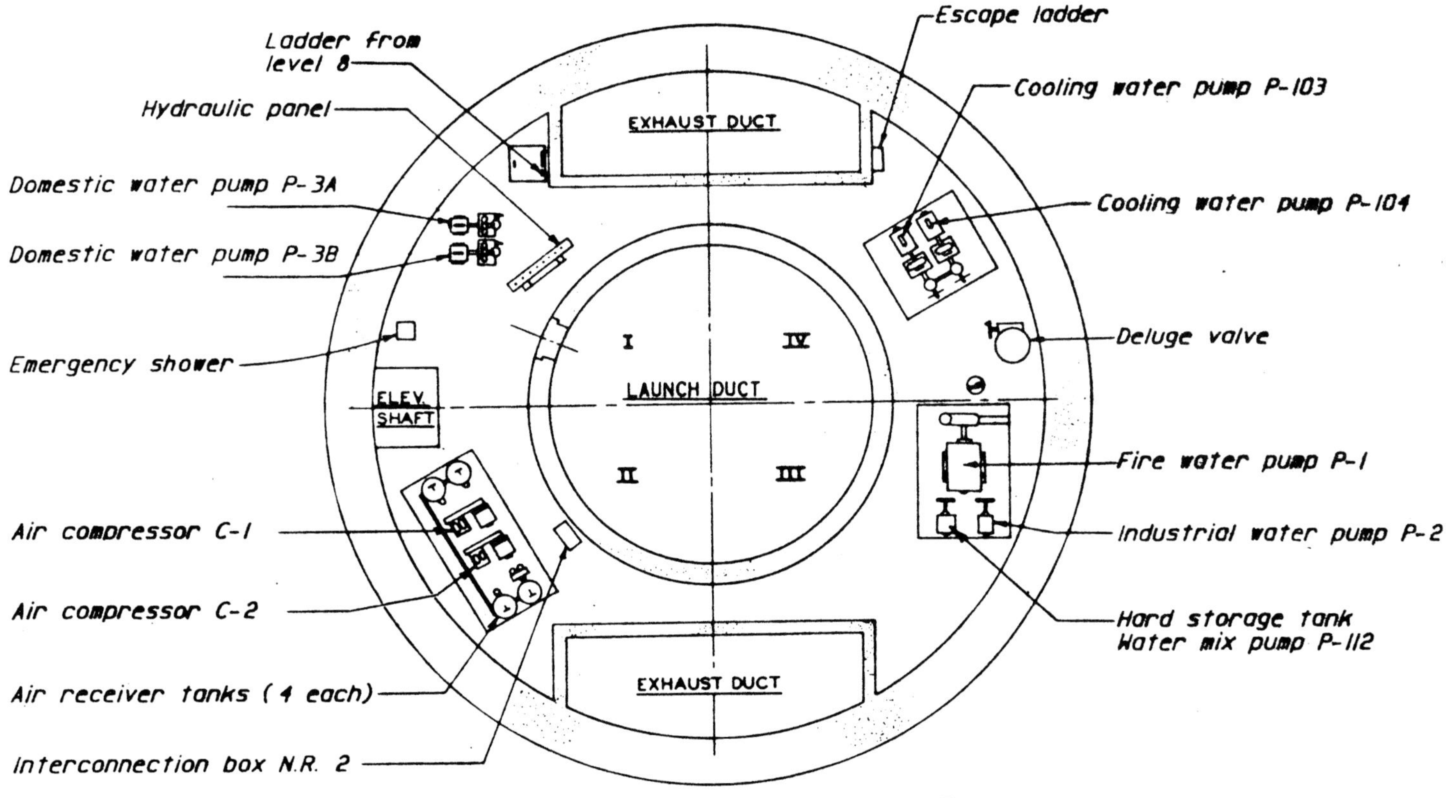

SILO EQUIPMENT AREA-LEVEL 7

▶ One of several emergency showers located in the silo.

▲ Cooling-water pumps circulated cooling water to the topside cooling towers. These pumps were not used after the installation of the modern topside coolers, except during plumbing shut-off.

these pumps would be closed.

In a plumbing shut-off scenario, the domestic water system would be cut off and the crew would be limited to the water in the pressure tank on LCC level 1. Pump P-2 would then be used to pressurize the industrial water system, utilizing the 100,000-gallon (380,000 l) tank as a supply.

Fire pump P-1 was driven by a 60-horsepower electric motor and was plumbed in parallel with pump P-2. P-1 would not start unless the demand for fire suppression water became so great that P-2 could not maintain sufficient pressure.[25] P-1 could deliver an astonishing 800 gallons (3000 l) per minute against 200 feet (61 m) of head pressure.[26]

Cooling water pumps P-103 and P-104 circulated water to the topside cooling towers and then returned it to the chiller units. When the underground chillers were replaced with mod-

▲ Fire pump P-1. At 60 horsepower, it was by far the largest electric motor on the complex.

◀ The industrial water pump P-2 (right) operated only during plumbing shut-off. Fire pump P-1 (at rear) started only if P-2 could not maintain adequate pressure. P-112 (left), a cold-water circulating pump, kept water in the silo tank in the correct temperature range.

◀ Air compressor unit. Starting-air for the diesel generator was supplied from the two tanks at right. The other tanks supplied air for the work platform locking pins and "utility air" for a variety of purposes.

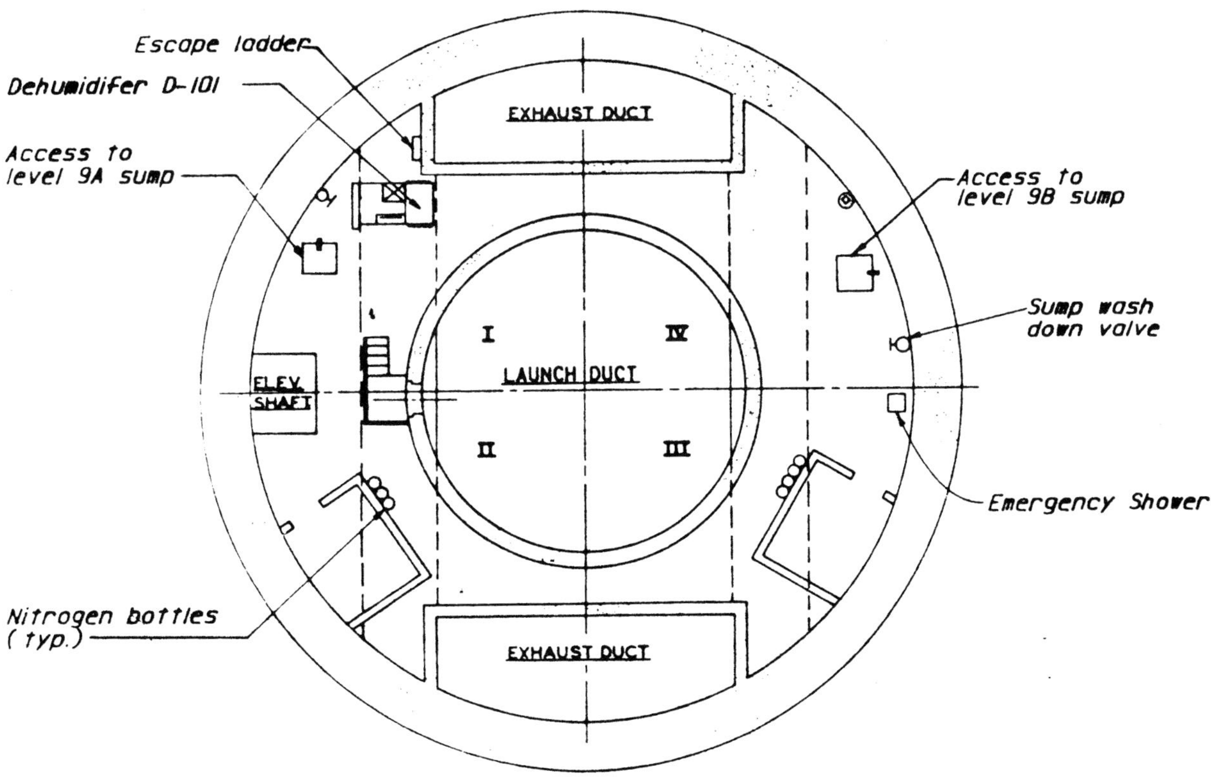

SILO EQUIPMENT AREA - LEVEL 8

TMM ARCHIVES

▲ The oxidizer pump room. Used in PTS operations for loading and unloading oxidizer on the missile.

USAF PHOTO, TMM ARCHIVES.

▲ The launch duct dehumidifier kept the relative humidity inside the launch duct at or below 30 percent.

ern topside units, these pumps were no longer required except in plumbing shut-off mode when they would be used to circulate cooling water through the silo water tank.[27]

P-112 circulated chilled water through the silo tank to keep the water at or below 73° F (23° C).[28]

Also on level 7 was the air compressor system that provided compressed air to the pneumatic systems on the complex. Air stored in two tanks on the left side was routed through air-drying canisters and used to operate several systems including the locking pins for the launch duct work platforms and the plumbing shut-off plug valves. Air from two tanks on the right side was used to start the diesel generator.[29]

Level 7 also provided access to the launch duct at the level of the stage one engine.

Silo Equipment Level 8

Loading propellant onto the missile was accomplished by gravity. Offloading was accomplished with pumps installed in separate small stainless steel rooms known as the fuel and oxidizer pump rooms. Both were located on level 8. As originally designed, the offloading pumps were permanently installed in each pump room. After emergency offloading capability was abandoned (along with the topside dump tanks) in 1979, the pumps were removed, and were brought to the complex only as required.

▶ The "diving board" (at upper right) on level 8 as seen from level 7. Note flame deflector. This massive concrete wedge directed missile exhaust into two vertical shafts.

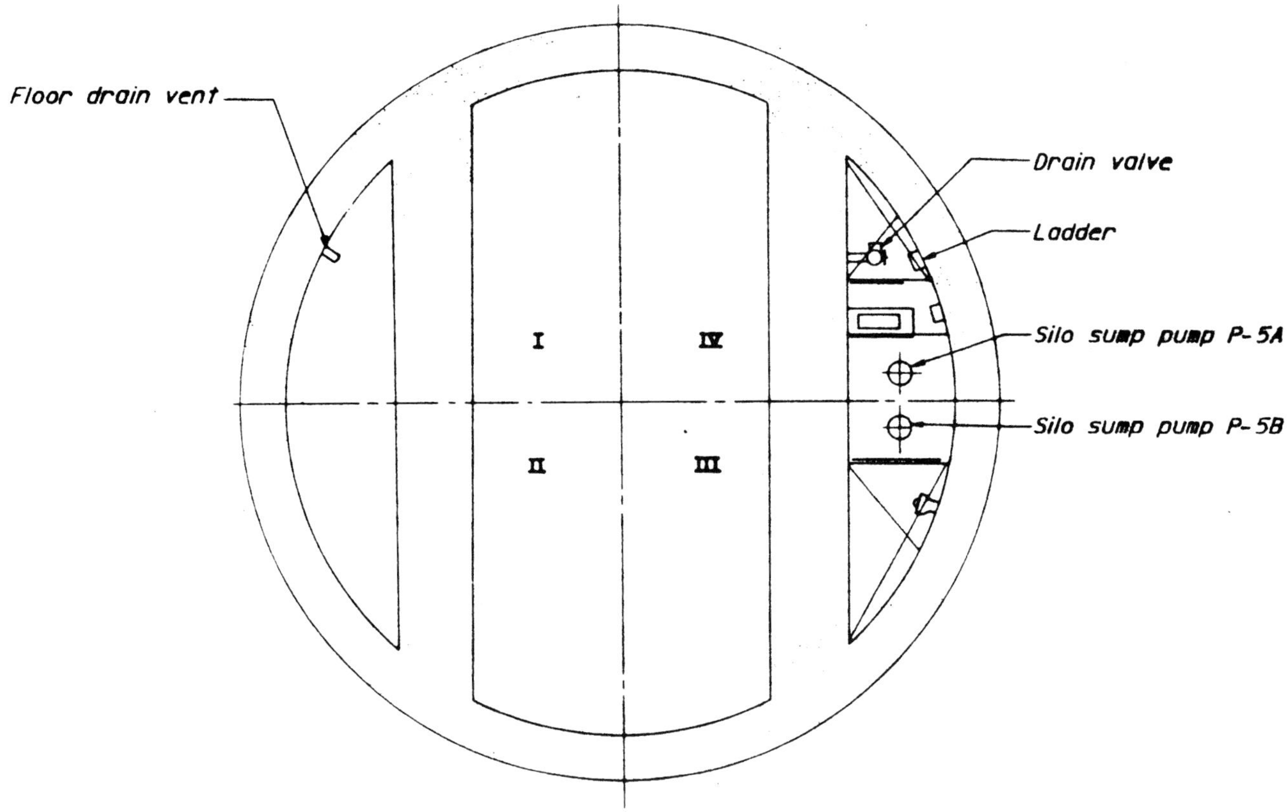

SILO EQUIPMENT AREA-LEVEL 9

TMM ARCHIVES.

▶ A view into the east sump. One of two sump pumps is visible at lower left.

◀ In spite of continuing efforts to plug leaks, water found its way into many silos. Water seeping in around this electrical grounding lug formed a cave-like formation of minerals on the silo wall.

USAF PHOTO, TMM ARCHIVES.

A launch duct access door on level 8 opened onto a small platform casually referred to as the "diving board." There were no work platforms at this level. A short ladder led down from the diving board to the top of the flame deflector. From there, step-irons on each side of the deflector led down to level 9.

Silo Equipment Level 9

Level 9 was the bottom of the launch duct. Here, the massive concrete wedge of the W would have directed the exhaust of the missile into two shafts that led up along opposite sides of the silo to vents on the surface.

Level 9 was subdivided into three separate areas. Levels 9A and 9B were sumps on opposite sides of the W, and were accessible only by way of hatches in the floor of level 8. Levels 9A and B served as drain points for all of the floor drains and emergency showers in the silo equipment area, as well as for the W. While infiltration of ground water was not a significant issue at most sites around Tucson, that was not the case at Little Rock and Wichita where the silo sump pumps ran almost continuously in some cases. Exactly how water found its way through four feet of concrete and a quarter inch of steel plating is not entirely clear, but it likely seeped in around pipes penetrating the silo wall, or perhaps through cracks or bad welds in the steel plating.

Level 9A contained no equipment but was connected to level 9B by a ten-inch (25 cm) diameter pipe that equalized the level of water in both sumps. Level 9B contained two pumps that pumped water to the surface, discharging it at grade-level through drains east of the fuel hardstand or, prior to 1979, into underground dump tanks. While the capability existed to pump water from the sumps, had the water been contaminated with propellant or other chemicals, federal law would have prohibited dumping it on the surface even if it had been "neutralized." Contaminated water would have needed to be pumped out and stored in tanks or trucks for proper disposal.[30]

▼ Hatches in the floor of level 8 opened into the level 9 sumps.

USAF PHOTO, TMM ARCHIVES.

▲ An "up the kilt" view of the missile as seen from the flame deflector. Note thrust mount ring and water deluge nozzles. With all work platforms stowed, this missile was good to go.

1 Ralph M. Parsons Company, Final Design Calculations, Structural, Volume 1 Section 4D, Launch Silo *Launch Duct: Relative Movement Between Duct and Silo,* sheets 1 through 9, October 1961

2 TO.21M-LGM25C-1, 1-32

3 00-ALCP-144-4 *Titan II Propellant Hazard Management Guide,* (Air Force Headquarters Ogden Air Logistics Center, 1983), A3-6

4 TO.21M-LGM25C-1, 1-46

5 Ralph M.Parsons Company, *Final Report Silo Door System Test Program,* WS 107 A-2 Titan II, 1962, 4-8, 4-9

6 SAC-CEM 35R-1-371-4, *Sensing, Warning, and Blast Protection,* 16 Feb 1976, Change 19, 6 Jan 1984, xxi-xxii

7 *Weapon System Familiarization,* Course OZR1821F/3121F, Technical Training, Shepard AFB, January 1965, 37

8 *Missile Systems,* Missile Launch/Missile Officer (LGM-25), Shepard AFB, May 1967, 68-72. TMM archives reference 20254.1

9 Ibid., 68-75, TMM reference 20254.2

10 TO.21M-LGM25C-1, 1-16

11 SAC-CEM 35R-1-351-4, *Power Generation and Distribution,* 1WD-12

12 Ibid. Typically, superchargers are driven from the crankshaft, while turbochargers are driven by the engine's exhaust stream.

13 SAC-CEM 21-SM68B-2-21-4, *Power Generation and Distribution,* 2-Feb-1979, Change 13, 1-May-1985, 1-9

14 TO.21M-LGM25C-1, 1-16

15 Ibid., 1-8

16 00-ALCP-144-4, A3-28, A3-29

17 Smith, interview

18 U.S. Army Corps of Engineers, Titan II Dismantlement, McConnell AFB, Sheet M-19

19 21-SM68B-2-21-4, 1-12

20 00-ALCP-144-4, A3-17

21 Ibid., A3-17

22 *Weapon System Familiarization,* Course OZR1821F/3121F, Technical Training, Shepard AFB, January 1965, 136

23 Ibid.

24 21-SM68B-2-21-4, 1-109—1-110

25 00-ALCP-144-4, A-3-18

26 *Weapon System Familiarization,* Course OZR1821F/3121F, Technical Training, Shepard AFB, January 1965, 33, 134

27 Ibid.

28 TO.21M-LGM25C-1, *Daily Shift Verification* checklist step 173

29 Sgt. Ralph Homke, MFT, USAF (Ret.), interview

30 00-ALCP-144-4, 4-57

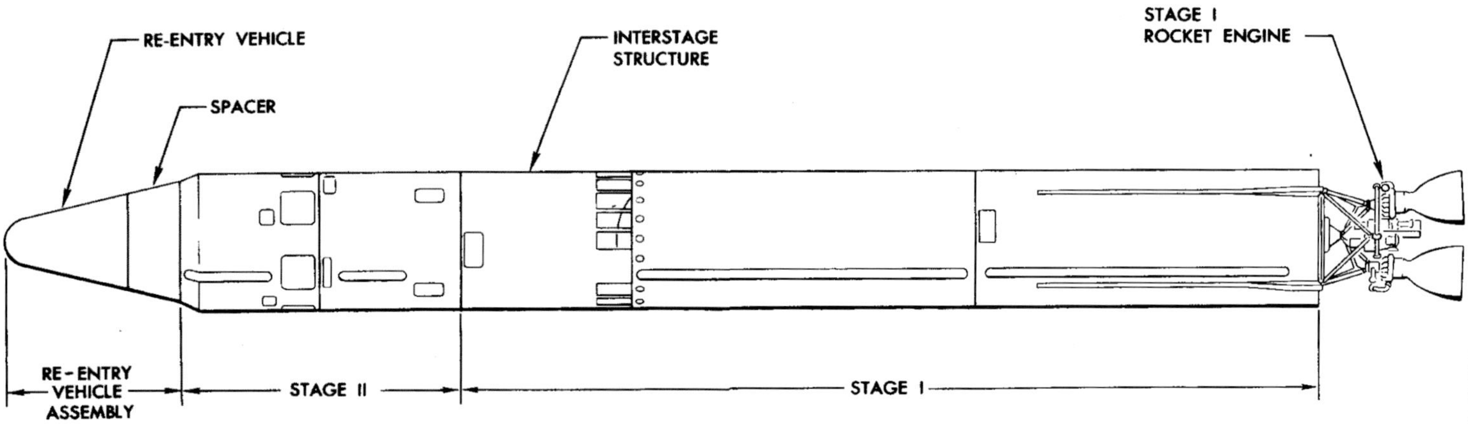

TMM ARCHIVES.

CHAPTER 7

THE MISSILE

Standing 103 feet (31 m) tall, with a range of over 6000 miles and topped with the largest thermonuclear bomb ever carried by a U.S. ICBM, Titan II was the blockbuster of the United States' nuclear deterrent. The main body of the missile, often referred to as the airframe, was 10 feet (3 m) in diameter and constructed from anodized aluminum, alloy type 2014-T6. The airframe was semi-monocoque (MON-a-coke) in design, meaning that the missile was self-supporting, and that its propellant tanks were an integral part of the airframe and not separate internal structures.[1] The thickness of the metal skin ranged from approximately .050 inches (1.27 mm) to .170 inches (4.3 mm) depending on the location. For comparison, a U.S. dime is .053 inches (1.35mm) thick, while a quarter is .068 inches (1.75 mm). Chemical milling was used extensively to reduce the thickness, and hence the weight, of the airframe. Titan II's weight, including the reentery vehicle (RV) and propellant, was nominally 330,000 pounds (150,000 kg)—a number made even more impressive when compared to the weight of the airframe and RV alone: just 23,000 pounds (10,400 kg).[2] Remove the RV and Titan II drops to a featherweight 15,000 pounds (6800 kg). Indeed, the missile was little more than a flying fuel tank. The first stage carried a combined total of about 25,500 gallons (96,500 l) of fuel and oxidizer, while the second stage carried about 6000 gallons (22,700 l).[3,4]

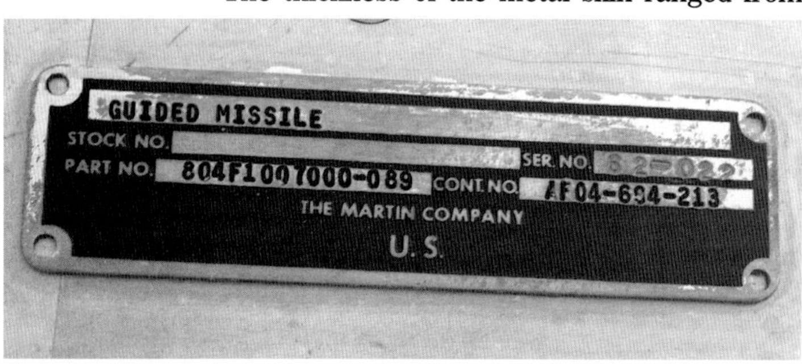

▲ Nameplate on underside of aft skirt.

The missile on display at the Titan Missile Museum, number N-10, was a training missile acquired from Sheppard AFB in Wichita Falls, Texas when the Titan IIs were retired, and was never loaded with propellant. While N-10's first stage is painted with black and white stripes, operational missiles were not so adorned. But all Titan II missiles, including N-10, were festooned with a multitude of labels on the airframe informing maintenance teams of the purpose or function of various panels and connectors.

Propellant—Making the Missile Fly

Throwing a 7600-pound (3450 kg) warhead a distance of more than 6000 statute miles (9600 km) requires tremendous energy. In Titan II that energy came from two very dangerous, toxic

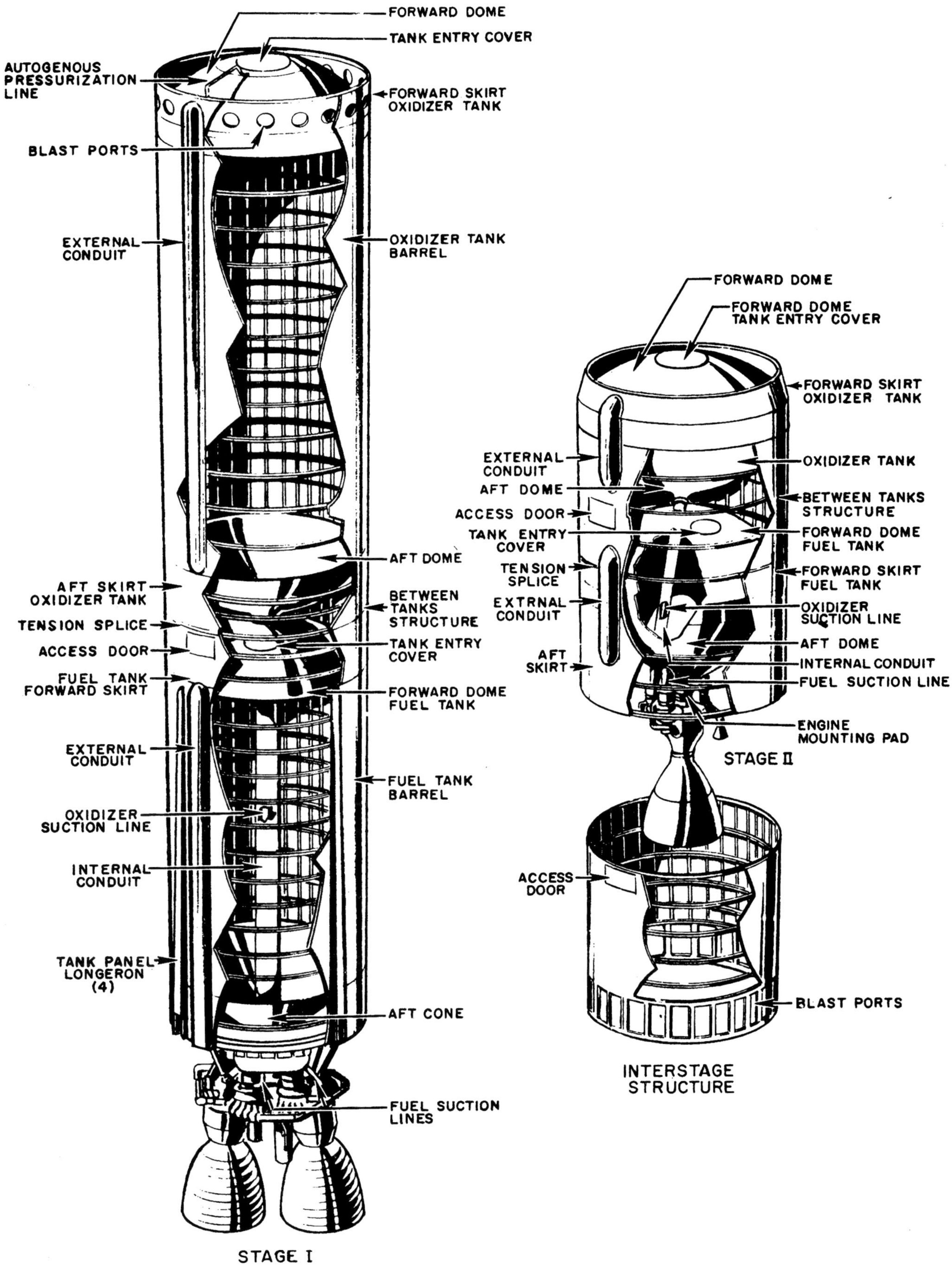

MISSILE AIRFRAME CONFIGURATION

Insertion

▼ The missile was transported to the complex by truck and lowered into the silo with a large crane. Stage one was installed first, followed by the stage two. The propellant was then loaded, and finally the RV was installed. This sequence of photos was assembled from two separate installations.

USAF PHOTOS, TMM ARCHIVES.

and volatile chemicals: a fuel and an oxidizer.

In your car the fuel typically used is gasoline, and the oxidizer is oxygen from the air. Gasoline was, in fact, used as a fuel in some early rockets and its chemical cousin, kerosene, was used in Atlas and Titan I. The oxidizer part is a little more complicated. Many rockets and missiles, including Titan II, fly beyond the Earth's atmosphere and therefore cannot depend on oxygen in the air to facilitate combustion. Indeed, most rockets and missiles carry their own supply of oxygen.

Titan II used a rocket fuel called Aerozine 50, a 50/50 blend of the chemicals hydrazine (N_2H_4), and unsymmetrical dimethyl hydrazine ($(CH_3)_2N_2H_2$), and an oxidizer, nitrogen tetroxide, (N_2O_4). Together, these two chemicals were called the propellant.

The huge advantage of these chemicals over kerosene and liquid oxygen was that they were stable enough to be stored onboard the missile for years at a time, waiting to go at a moment's notice. While Atlas and Titan I required precious tens of minutes to be loaded, Titan II was always fully loaded and could be launched in just 58 seconds.

◄ The rocket fuel handler's clothing outfit, or RFHCO, worn to protect against the highly toxic vapors of the propellant. The suits were not air-conditioned.

USAF PHOTO, TMM ARCHIVES.

But this huge advantage brought with it two huge disadvantages. First, Aerozine 50 and nitrogen tetroxide are exceedingly toxic and had to be handled with what amounted to moon suits. Those involved with propellant handling wore a special protective suit called a rocket fuel handler's clothing outfit (RFHCO, pronounced REFF-co). The suits were rubberized and were not air conditioned; wearing them in 100° F (38° C) weather was not a pleasant task. Still, it beat the alternative of being poisoned by fuel or oxidizer vapors.

Second, nitrogen tetroxide had to be kept at the correct temperature in order to remain stable. If it got too warm, it would begin to boil. But an equally important consideration was maintaining the propper specific impulse—the amount of power produced by the engine every second. Since the mass of a liquid changes with temperature engineers had to agree on a standard temperature so as to maintain the correct mass/volume relationship of the propellant. Engineers settled on 60° F (15.5° C) as a mass/volume standard with adaqaute headroom to keep the oxidizer from boiling. The tolerance was plus or minus two degrees. Both oxidizer and fuel—along with the entire missile and launch duct—were kept at this critical temperature.

There is a third point regarding the propellant that is an advantage and a disadvantage all rolled up together, and that is that the propellant was hypergolic, meaning that the fuel and oxidizer would ignite spontaneously when mixed. The advantage was that there was no complicated ignition system to fail. This disadvantage was that if the fuel and oxidizer mixed accidentally, as from a leak in the missile or some kind of in-silo accident, the result could be catastrophic.

Loading and unloading Titan II's propellant required a great deal of specialized equipment, the most conspicuous of which were the holding trailers (large trailer-mounted tanks) and conditioning trailers (truck-like heating/cooling units). The process of loading or unloading propellant, together with all of the equipment required to handle the volatile chemicals, was referred to collectively as the propellant transfer system (PTS). Prior to loading the propellant into the missile it was brought to the correct temperature with the conditioning trailers. Once conditioned, the propellant was gravity fed into the missile in the silo. The missile's propellant tanks were then pressurized with nitrogen to prevent any reactions with oxygen or water vapor.

PTS operations were complex, labor intensive, time consuming and dangerous. Whether in the silo or on the surface, spills of fuel or oxidizer represented a potentially life-threatening event, not only for those in the immediate vicinity of the spill, but also for the crew in the control center and anyone downwind of the complex. Aerozine 50 is extremely volatile and in sufficient concentrations will react violently with a wide variety of common substances such as paints, oil, grease and solvents, and even rust. And as mentioned earlier, oxidizer created large volumes of highly toxic and corrosive vapors. Fortunately, propellant spills were an extremely rare occurrence.

The propellant had no "shelf life," so in theory, once loaded it could remain on the missile forever. In practice, however, the chemicals were corrosive enough that gaskets and seals might begin to leak after a number of years. In that case the propellant would need to be off-loaded, then reloaded once repairs were completed.

Stage One

Rocket scientists and engineers are loath to talk about the power of a rocket engine in terms

▼ The LR-87-AJ-5 stage-one rocket engine. It had no throttle control or "off switch" and burned at full power until the propellant ran out. Its thrust was roughly equivalent to that of two Boeing 747s.

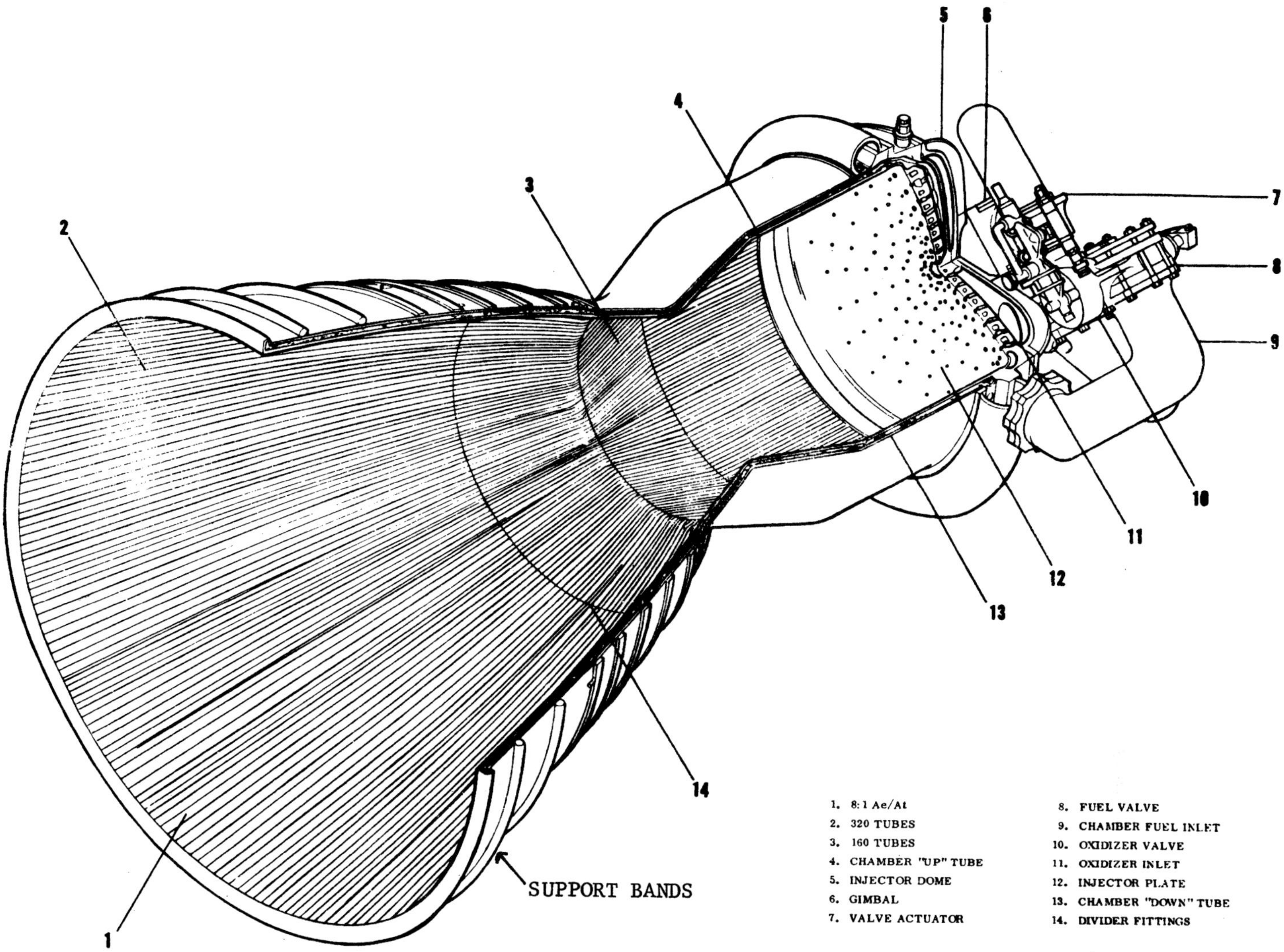

▲ Detail of stage-one thrust chamber assembly.

of horsepower (with good reason), preferring always to speak of pounds of thrust. The stage-one engine had two thrust chambers that together developed 430,000 pounds (195,000 kg) of thrust: just a bit less than two 747s running at full throttle.[5] Yet for most people, horsepower is the most familiar unit of measure with respect to engines and motors. By definition, horsepower is simply an expression of a certain amount of work done in a specific length of time, and so it is possible to express the power of the stage-one engine in horsepower, at least at any one instant in time. Archival video of Titan II launches provides the data required to make the calculation. As the missile clears the silo it can be seen to move its length, 103 feet (31.4 m), in three seconds.[6] And as the weight of the missile at this point in the flight is known with a reasonable degree of confidence (very close to 330,000 pounds (149,600 kg)), the calculation is straighforward. Right out of the silo, Titan II develops about 20,000 horsepower. After that point, things get much more complicated because the missile's speed and weight change continuously.

The engine may look complex, but its design was straightforward. In simplest terms, fuel and oxidizer were mixed and ignited in a small cavity call a thrust chamber. The resulting combustion generated huge volumes of very hot gas. This gas forced its way out of the chamber through a constriction at one end, exiting at supersonic speed. The flow of exhaust gas in one direction pushed the missile in the other direction, fulfilling Newton's Third Law: For every action there is an equal and opposite reaction.

A gas-powered turbine on each subassembly, initially started with a device called a start cartridge, drove propellant pumps on the engine. The start cartridges were essentially miniature solid propellant rockets. When ignited, their exhaust was blown into the turbines, providing a kick-start and accelerating the turbines from zero to 24,000 RPM in just two seconds. If 24,000 RPM sounds brutal, it was. But the turbines and pumps only had to hold together for 2½ minutes. Remember, this is a one-way trip. Once the turbines were in motion, small quantities of fuel and oxidizer were drawn into a gas generator where they ignited by hypergolic action, providing a flow of high-pressure gas that kept them spinning. The fuel and oxidizer

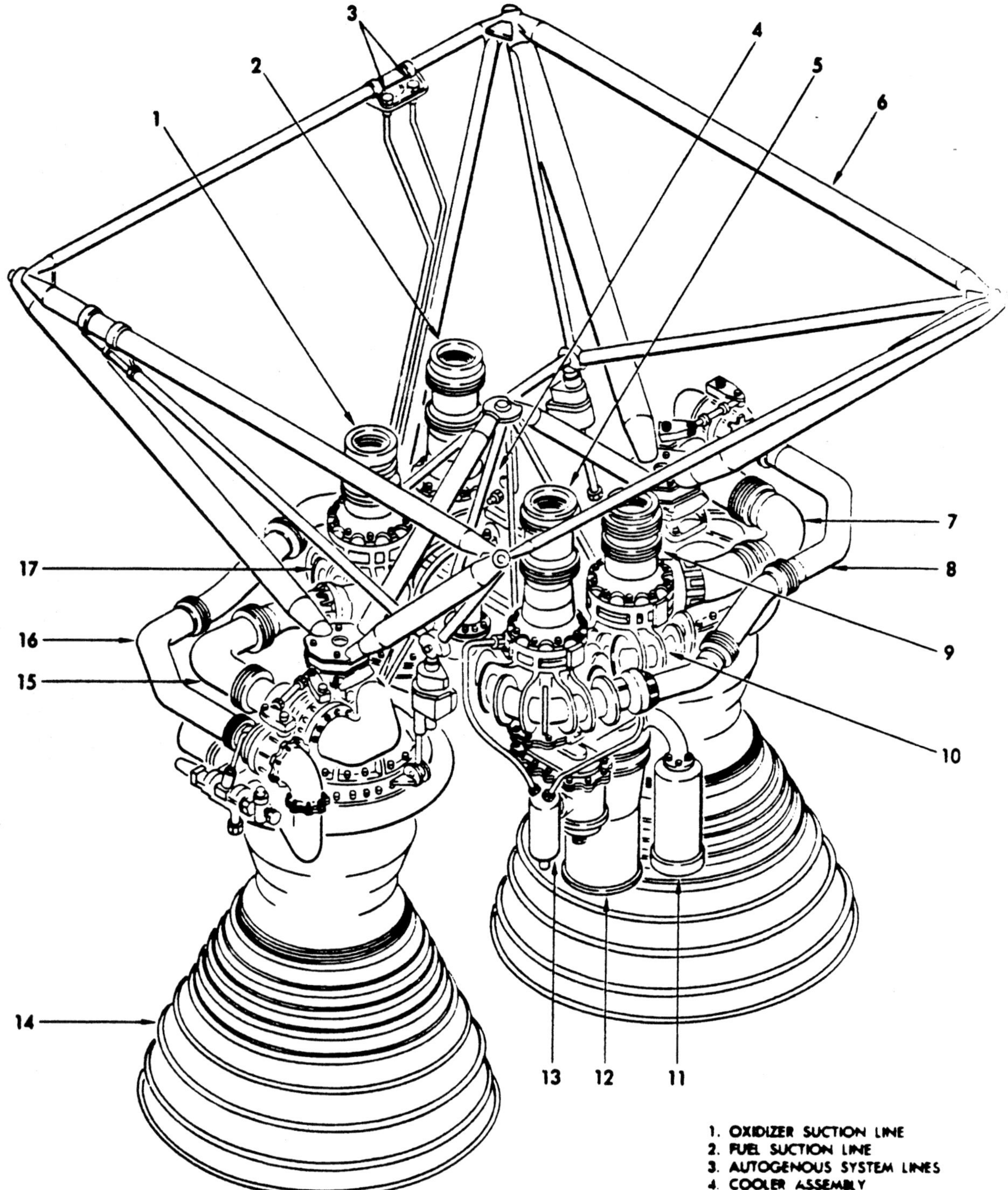

1. OXIDIZER SUCTION LINE
2. FUEL SUCTION LINE
3. AUTOGENOUS SYSTEM LINES
4. COOLER ASSEMBLY
5. FUEL SUCTION LINE
6. ENGINE FRAME
7. OXIDIZER DISCHARGE LINE
8. FUEL DISCHARGE LINE
9. OXIDIZER SUCTION LINE
10. TURBOPUMP ASSEMBLY
11. GAS PRESSURE GENERATOR
12. EXHAUST STACK
13. OIL COOLER
14. THRUST CHAMBER ASSEMBLY
15. OXIDIZER DISCHARGE LINE
16. FUEL DISCHARGE LINE
17. TURBOPUMP ASSEMBLY

Stage I Rocket Engine

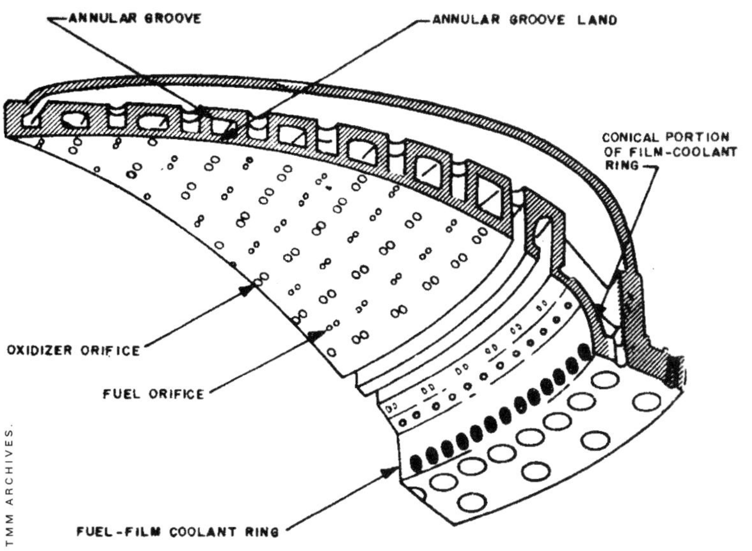

TMM ARCHIVES.

▲ Diagram of a portion of the injector plate.

pumps were connected to the turbines with a set of gears.

The pumps forced fuel and oxidizer through an injector plate at the top of each thrust chamber. The propellant squirted out through pairs of small holes in the plate and became a mist. Hypergolic combustion was instantaneous.

Once in flight, signals from the missile's guidance computer operated hydraulic actuators that moved the thrust chambers up and down and from side to side as required to keep the missile on course.[7] The thrust chambers could be moved independently, permitting the missile to be controlled in pitch (moving the nose of the missile up or down), yaw (moving the nose from side to side) and roll (rolling the missile about its long axis).

The first-stage engine burned for just 2½ minutes but consumed a combined total of about 25,500 gallons (96,500 l) of fuel and oxidizer—about 170 gallons (640 l) per second. Stage one boosted the missile to an altitude of approximately 50 statute miles (80 km) and propelled it about 50 statute miles down range (toward the target).[8]

Radio communications and destruct packages were flown only on research-and-development and operational test flights. The inability to destroy or communicate with the missile might sound like a mistake, but there was a good reason: if the Air Force could send radio commands to the missile, the enemy could probably figure out how to jam the signals or send its own.

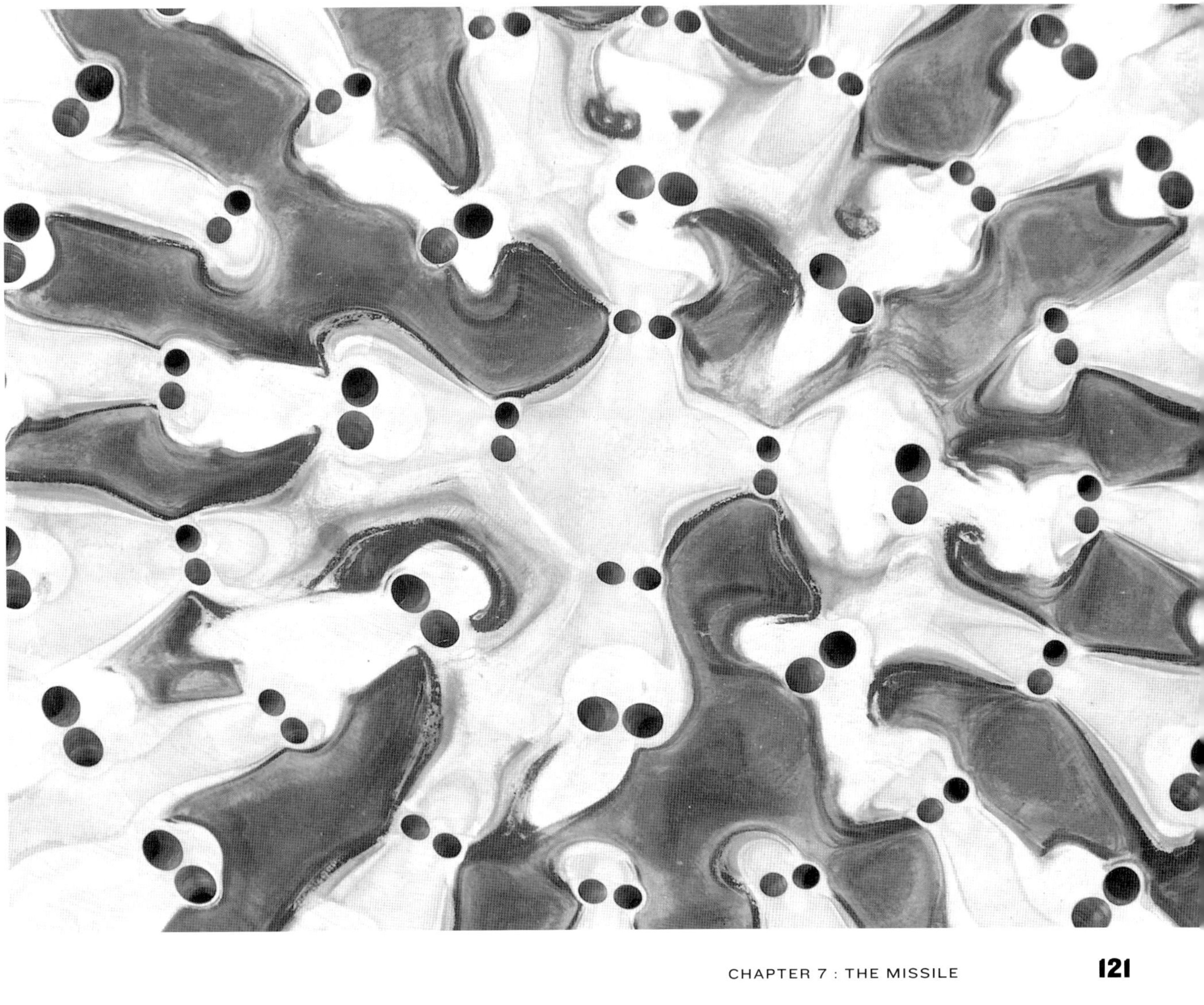

▶ Close-up of injector hole pattern. Hole pairs created impinging streams resulting in a mist. The larger holes were for oxidizer, the smaller ones for fuel. Dark areas are corrosion due to weathering. This plate was found in a dumpster full of engine parts at the Aircraft Maintenance and Regeneration Group, Tucson's famous aircraft "boneyard."

◀ Rare declassified photo of the butterfly valve locking assembly. The unit was inserted into a stage-one oxidizer intake pipe. The valve remained locked until a six-letter unlock code was entered in the butterfly valve lock control on LCC level 2.

USAF PHOTO, TMM ARCHIVES.

Butterfly Valves

There were four propellant intake lines on the stage-one engine (one fuel and one oxidizer for each subassembly), each fitted with a pre-valve, often referred to as a "butterfly" valve (in spite of the fact that their operation didn't really mimic that of a butterfly at all). The pre-valve on one of the stage-one oxidizer lines had a built-in locking device operated by a sophisticated digital electronics package. This unit compared its internal code with a code received from the butterfly valve coded-switch control in the LCC. Unless the codes matched, the valve would not unlock. Attempts to disassemble or manually unlock the valve were met with the discharge of a small pyrotechnic device that jammed the valve in the closed position.[9,10]

Thrust Mount and Shock Isolation

The missile rested on a large, 13-ton (11.8 metric ton) donut-shaped platform called the thrust mount, located just above the level 7 launch duct access door. The missile was attached to the thrust mount with four hold-down bolts. These bolts were secured with explosive nuts at both ends. When the engine

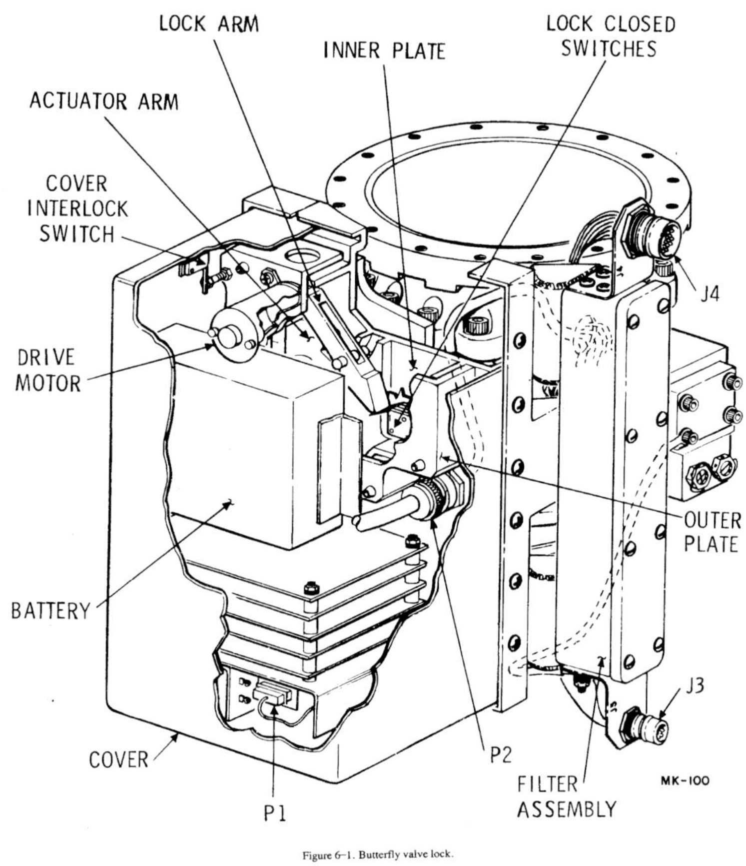

Figure 6-1. Butterfly valve lock.

◀ Declassified schematics of butterfly valve assembly.

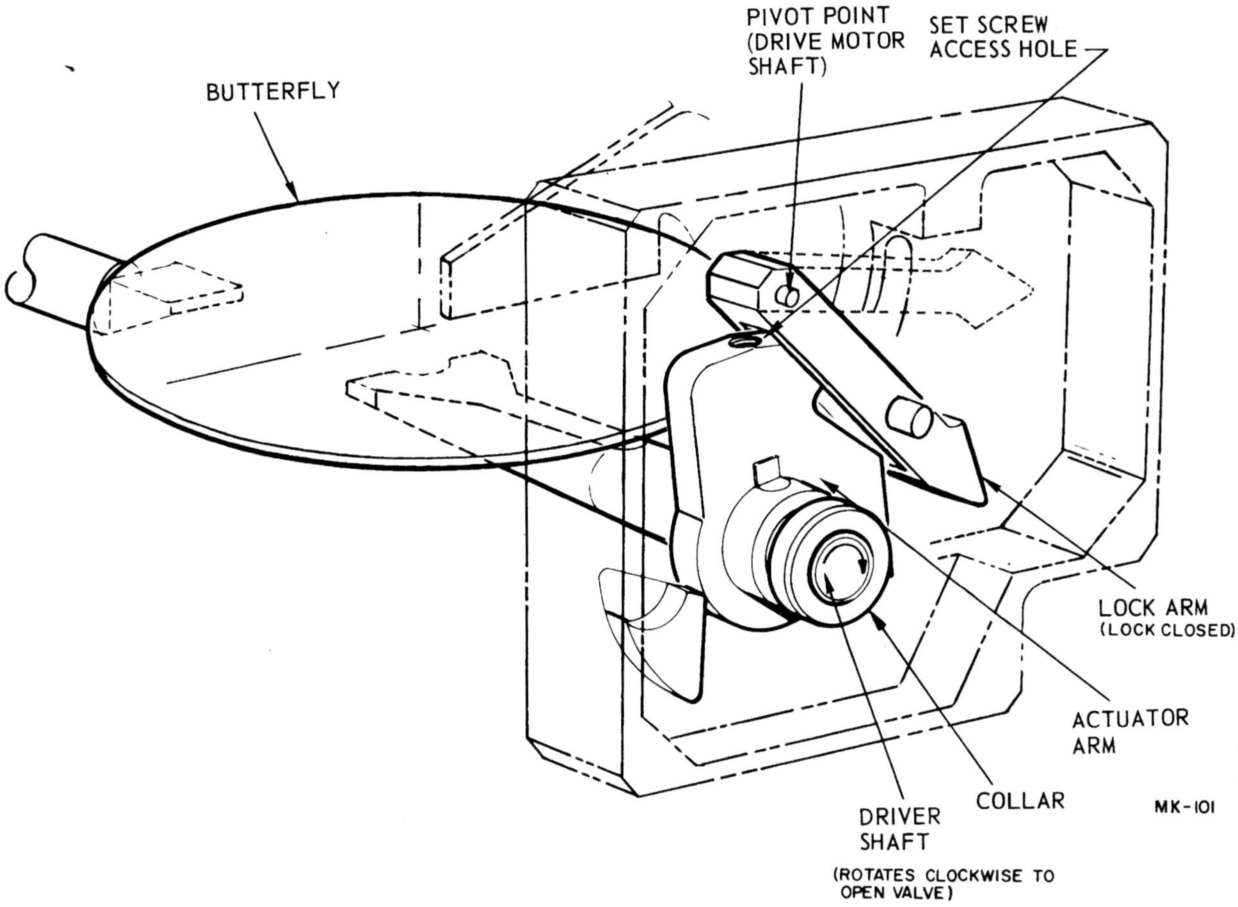

Figure 6–2. Locking mechanism, butterfly valve lock.

started, pressure sensors waited for the engine to develop 77 percent of its rated thrust. Extensive tests of the engine revealed that if the engine reached this level of thrust, there was a 99 percent probability it would reach full power. When the 77 percent thrust level was detected (typically within two seconds of start-up), timers counted down a total of 1.8 seconds to ensure that engine thrust was both stable and symmetrical. If all indicators were good when the timers timed out, a signal was sent to explode the hold-down bolts, and the missile was free to lift off.[11]

The thrust mount served two important functions: it provided shock isolation for the missile and a stable platform from which to launch. The thrust mount was suspended from the launch duct wall by four 30-foot (9 m) springs. Under normal conditions the thrust mount was free to move up and down and from side to side in response to a shockwave or earthquake, functioning very much like the shock absorbers on a car. This elastic state of the thrust mount was said to be "hard" because it could protect the missile from shock. Just prior to lift-off, the thrust mount was locked into position. The locked condition was called "soft" since the missile was then vulnerable to shock.

Stage Two

The engine on stage two, also made by Aerojet, worked very much like the stage-one engine. Its single thrust chamber produced 100,000 pounds (25,000 kg) of thrust. With only one thrust chamber, pitch and yaw control would be simple enough, but controlling roll would be impossible. Since roll control was essential, engineers made clever use of the exhaust from the propellant turbopumps. The exhaust was routed to a roll nozzle, forming what amounted to a thrust chamber in miniature, which could be moved perpendicular to the main engine.[12]

When stage-one propellant was exhausted sensors detected a decreased level of thrust and sent a signal to ignite stage two. The second stage engine fired up while it was still connected to the first stage. Explosive bolts joining the first and second stages then fired, separating the two stages. This technique guaranteed continuous acceleration,

▲ The thrust mount ring provided both shock isolation and a stable platform from which to lift off.

preventing the now-weightless liquid propellant from drifting away from the intake pipes, which would cause the engine to flame-out or fail to start.[13]

The stage-two engine burned for about three minutes and took the missile to an altitude of about 200 statute miles (320 km).[14] Unlike the stage-one engine, stage two had an off switch. When the guidance system determined that the missile was traveling at the correct speed and in the right direction to hit the target, the engine was shut down. Simultaneously, two small solid-propellant vernier motors ignited to fine-tune the speed and direction. The vernier motors were metal spheres about 14 inches (35 cm) in diameter fitted with small conical nozzles. The vernier motors burned for 20 seconds. Once ignited, solid-propellant motors cannot be shut off, so when vernier thrust was no longer required, an explosive charge blew off the nozzles, without which, thrust disappeared.

After vernier thrust cut-off the nuclear

▲ The missile was attached to the thrust mount by bolts with explosive nuts on each end. If one nut failed to explode, the bolt either stayed behind or left with the missile, but the missile took off—no matter what.

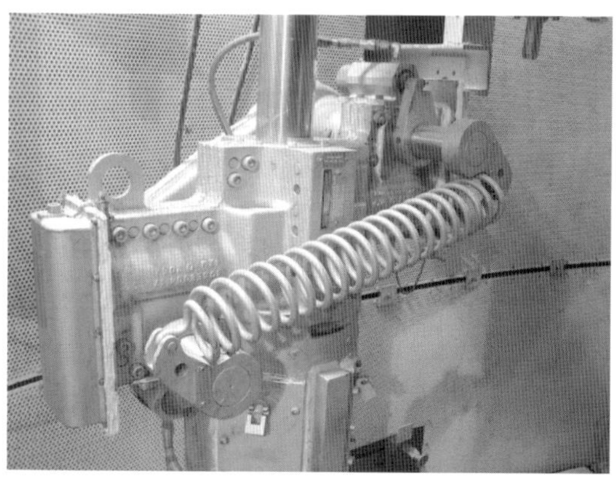

◀ Seconds before ignition large spring assemblies slam closed to lock the thrust mount in place.

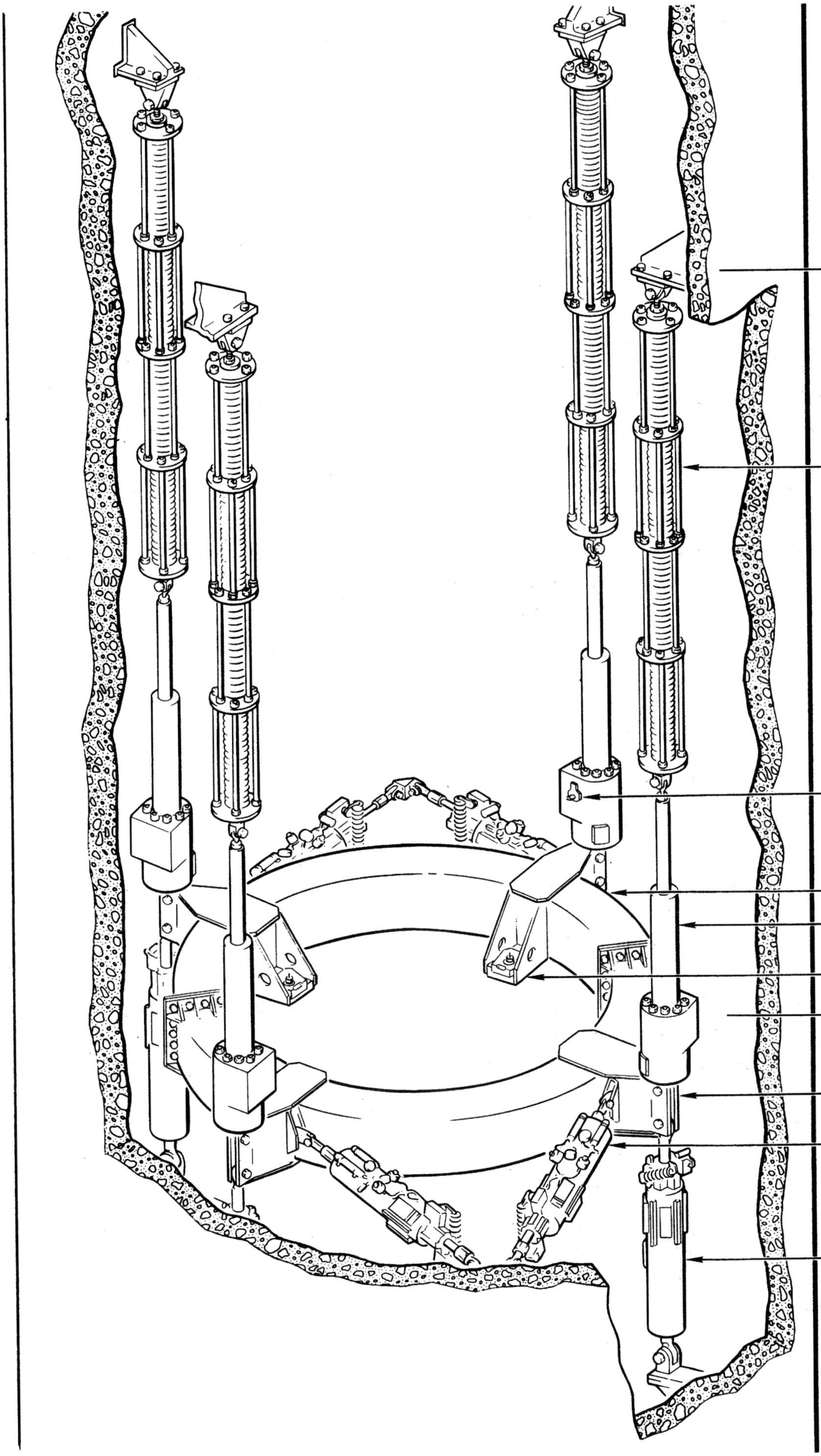

TMM ARCHIVES.

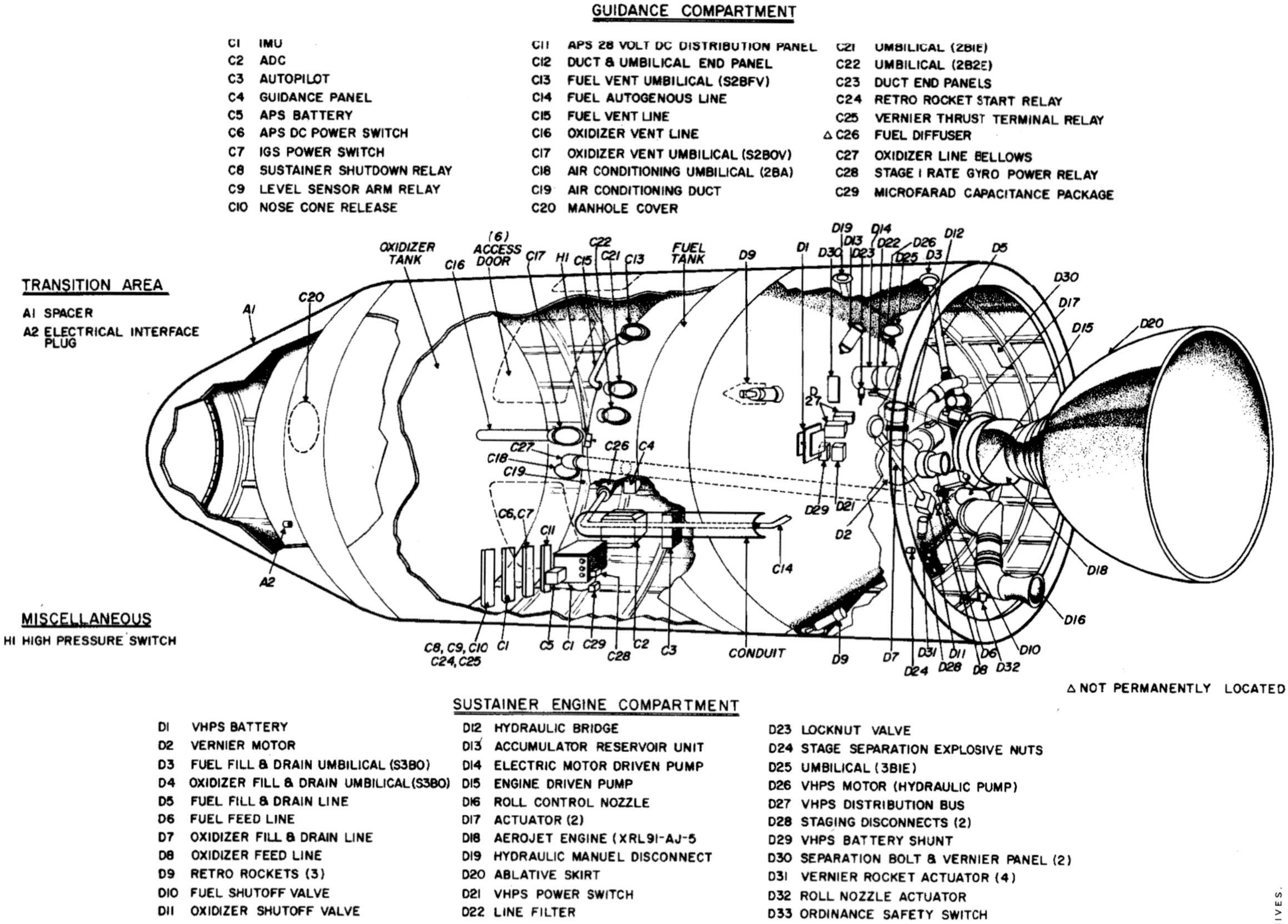

Figure 1, Stage II Component Location

TMM ARCHIVES.

◄ LR-91-AJ-5 stage-two rocket engine.

warhead was pre-armed and a Primacord explosive charge was fired, cleanly slicing the metal adapter connecting the RV to the second stage.[15] At that instant, the RV was free of the missile and on its own. An instant later, two small pitch rockets on the stage two were fired sequentially.[16] The first pitch rocket had the effect of backing stage two away from the RV (to prevent a collision between the two) and rotating (or pitching) stage two radially. The second pitch rocket then fired to stop the rotation.

What happened next depended on the time frame under discussion. Prior to 1975 decoys were ejected and two large translation rockets attached to stage two fired, pushing the stage-two airframe into a "specific orientation" with respect to the RV and its decoys. The RV, decoys and stage two airframe all reentered the atmosphere together.[17] When stage two began to break up due to heating from atmospheric drag, its fragments provided additional cover

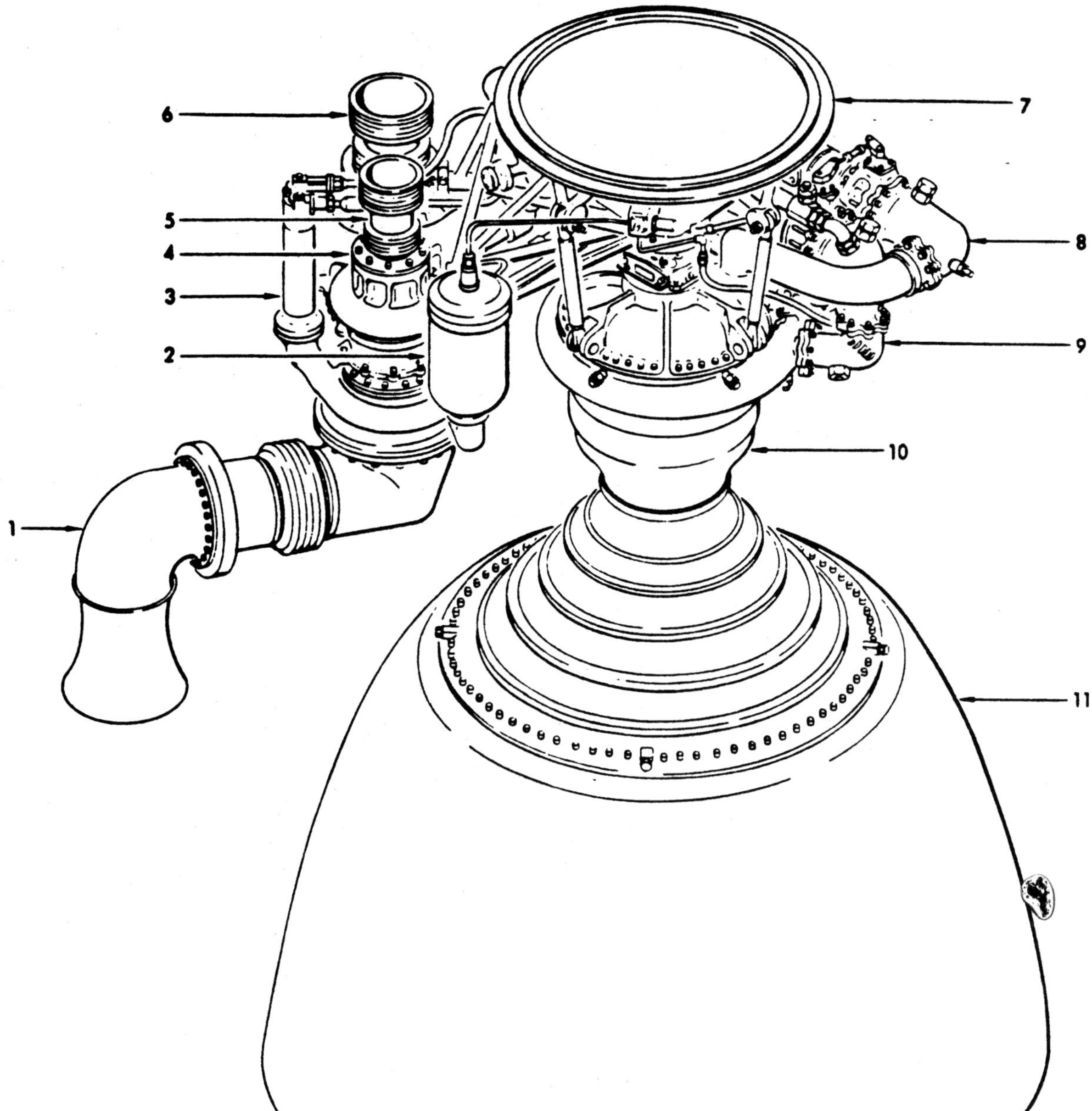

1. ROLL CONTROL ASSEMBLY
2. GAS PRESSURE GENERATOR
3. GAS GENERATOR
4. TURBOPUMP ASSEMBLY
5. FUEL SUCTION LINE
6. OXIDIZER SUCTION LINE
7. ENGINE FRAME
8. OXIDIZER DISCHARGE LINE
9. FUEL DISCHARGE LINE
10. THRUST CHAMBER ASSEMBLY
11. ABLATIVE SKIRT

Stage II Rocket Engine

▶ Thiokol solid-propellant vernier motor. Two verniers on stage two fine-tuned the velocity and attitude of the missile. Each motor developed about 500 pounds (230 kg) of thrust.

▶ One of two pitch-rockets used to back stage two away from the RV.

▶ One of two translation rockets side-mounted on Titan II until 1975. Each one delivered 5000 pounds (2300 kg) of thrust to orient stage two after RV release. These rockets were not required after removal of the decoy system and were subsequently deleted.

USAF PHOTO, TMM ARCHIVES.

▶ Location of missile guidance components aboard Titan II.

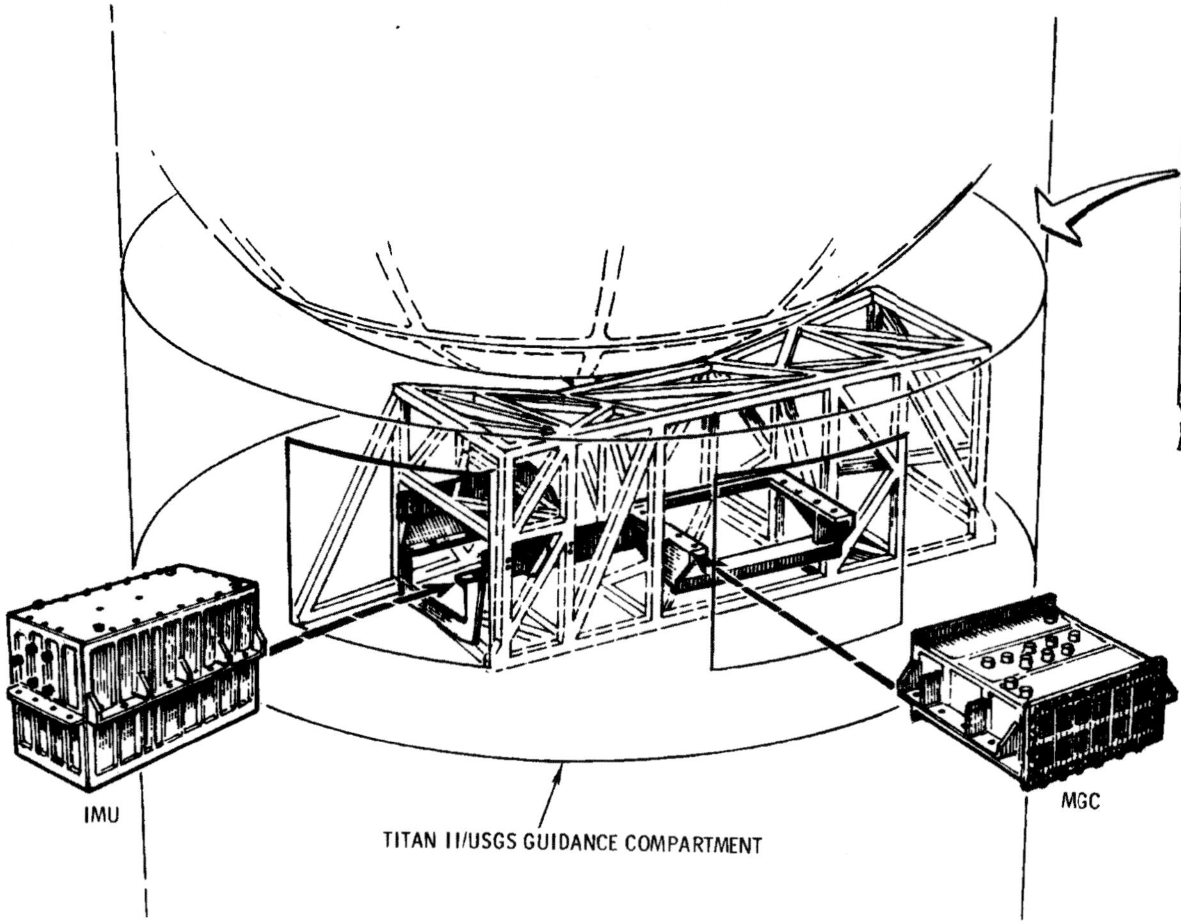

TMM ARCHIVES.

▶ The Carousel IV inertial measurement unit (IMU) tracked the missile's acceleration and attitude.

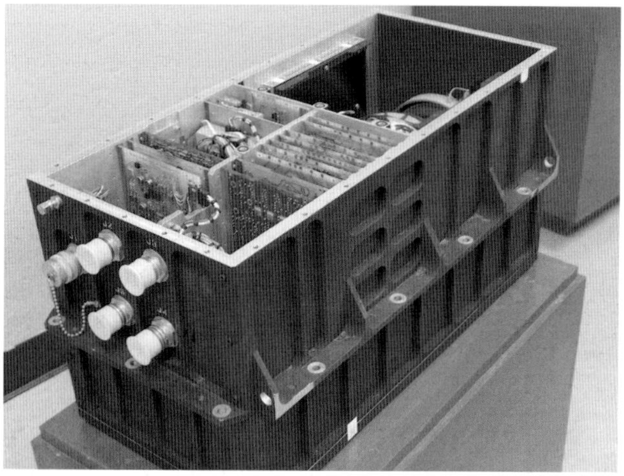

▲ Delco's Magic 352 guidance computer. Designed around 1970, it was 10 years ahead of its time. The computer had 16k of ferrite-core RAM, a fabric-like memory woven with thousands of tiny ferrite beads.

for the reentry vehicle.

The decoys were removed in 1975 because the anti-ballistic missile (ABM) systems that could have shot the RV out of the sky, and which the decoys were meant to distract, never materialized. ABMs were all but outlawed under the ABM treaty of 1972.[18] No decoys meant the translation rockets were no longer required, and they too were removed. After 1975 the aforementioned pitch rockets simply pushed stage two back and out of the way of the RV. Aerodynamic differences then pushed the two further apart as they reentered the atmosphere.

Guidance System

Titan II's guidance system consisted of three major components: the missile guidance computer (MGC), the inertial measurement unit (IMU) and the missile guidance-alignment checkout group (MGACG).[19]

The MGC and IMU were located on the missile itself in sealed, pressurized aluminum boxes between the fuel and oxidizer tanks inside stage two. The MGACG was located in the launch control center (LCC), and was used to program the MGC with the target trajectories.

The IMU sent information regarding the veocity and attitude of the missile to the MGC. (To understand attitude, consider the example of a car sliding on an icy roadway. While still moving straight ahead, the car may slip sideways so that the driver is looking at the

INERTIAL MEASUREMENT UNIT CHARACTERISTICS	
TYPE	Rotating XY platform All attitude Four gimbal
GYRO	Three AC651G Single-degree-of-freedom Gas Bearing Magnetic Suspension
ACCELEROMETER	Three AC653A Seismic mass
ELECTRONICS	Integrated digital and linear devices, high frequency switching control
COOLING	Radiant and Conductive
POWER	150 watts 28 Vdc input (nominal)
SIZE	22" x 12" x 10"
SURFACE AREA	8.5 ft^2
VOLUME	1.6 ft^3
WEIGHT	80 pounds

side of the road. It can be said that while the car's direction is correct, its attitude is not). The IMU was a collection of accelerometers and gyroscopes. The gyroscopes measured the attitude of the missile, while the accelerometers measured its speed. The IMU sent this information to the MGC, which in turn sent steering commands to the engine, thus keeping the missile on course.[20]

Titan II's original guidance computer was made by IBM under contract to AC Spark Plug, and was designed in 1959. Through the 1970s

MAGIC 352 COMPUTER CHARACTERISTICS	
COMPUTER TYPE	General purpose, binary, parallel, programmable, two's complement, fixed point
MEMORY TYPE	Random access, coincident current, toroidal core
MEMORY SIZE	16,384 words
WORD SIZE	24 bits, plus memory parity
MEMORY CYCLE	3.0 μs
ADDRESSING	Single address—direct, indirect, and relative
INTERRUPTS	8 priority levels
INSTRUCTIONS	57
EXECUTION SPEEDS	
• Add/Subtract	6.0 μs
• Multiply	30 μs
• Divide	36 μs
• Input/Output	6.0–27 μs
• Double Precision Add/Subtract	9.0 μs
INPUT/OUTPUT	Program controlled and direct memory access Serial and parallel channels
LOGIC	Transistor—Transistor, MSI
COOLING	Radiant and Conductive
POWER	CPU and Memory – 85 watts Total including I/O – 220 watts, 28Vdc
SIZE	16" x 20" x 9"
SURFACE AREA	9 ft^2
VOLUME	1.6 ft^3
WEIGHT	80 pounds
Computer including CPU, memory, power supply and all I/O modules.	

▶ Detail of the Carousel IV IMU gyroscope and accelerometer package.

TMM ARCHIVES.

◄ General Electric's Mark VI reentry vehicle carried the W-53 thermonuclear bomb. The cut out section seen in this image proves the RV is not functional.

spare parts for the unit became increasingly difficult to acquire, necessitating a decision to find a replacement.

As luck would have it Delco Electronics had developed a guidance system for use in Boeing's 747. The computer was called the Magic III and the inertial measurement unit was the Carousel IV. The system was later modified and rebranded as the Universal Space Guidance System (USGS) for use in the Titan III-C space launch system. The computer was now called the Magic 352. USGS was presented to the Air Force as replacement guidance system for Titan II.[21] Tests began in 1975, and installation of USGS aboard Titan IIs began in 1978. The changeover was completed fleet-wide by the end of 1979.[22]

A point of interest for those familiar with older computers: the Magic 352 had 16k of ferrite-core RAM and used both small and medium-scale integrated circuits. In speed and power, the 352 might be compared to the Commodore VIC-20, released in 1981, except the VIC-20 had only 5k of RAM.[23,24]

Reentry Vehicle

General Electric built Titan II's Mark VI reentry vehicle, the nose cone of the missile. The term "reentry vehicle" refers to the fact that it would fly out of the atmosphere and reenter it later on. It was the reentry vehicle (almost always shortened to RV) that carried Titan II's deadly payload, the W-53 hydrogen bomb, generally referred to as the warhead. The Air Force has never declassified the yield of the W-53, though numerous credible sources place the yield at nine megatons.[25,26] That is roughly 600 times the yield of the atomic bombs dropped on Japan near the end of WWII.

A great many details of the RV and bomb

▼ The 9-megaton W-53, sometimes referred to euphemistically as the "physics package."

COLLECTION OF THE NATIONAL ATOMIC MUSEUM, AUTHOR'S PHOTO

◀ An adapter section mated the RV to the missile. It contained a number of cylinders filled with metalized balls. Prior to reentry a Primacord charge opened metal panels on the adapter and a mechanism ejected the balls as decoys. In 1975 the decoy system was abandoned, but the cylinders in the adapter remained.

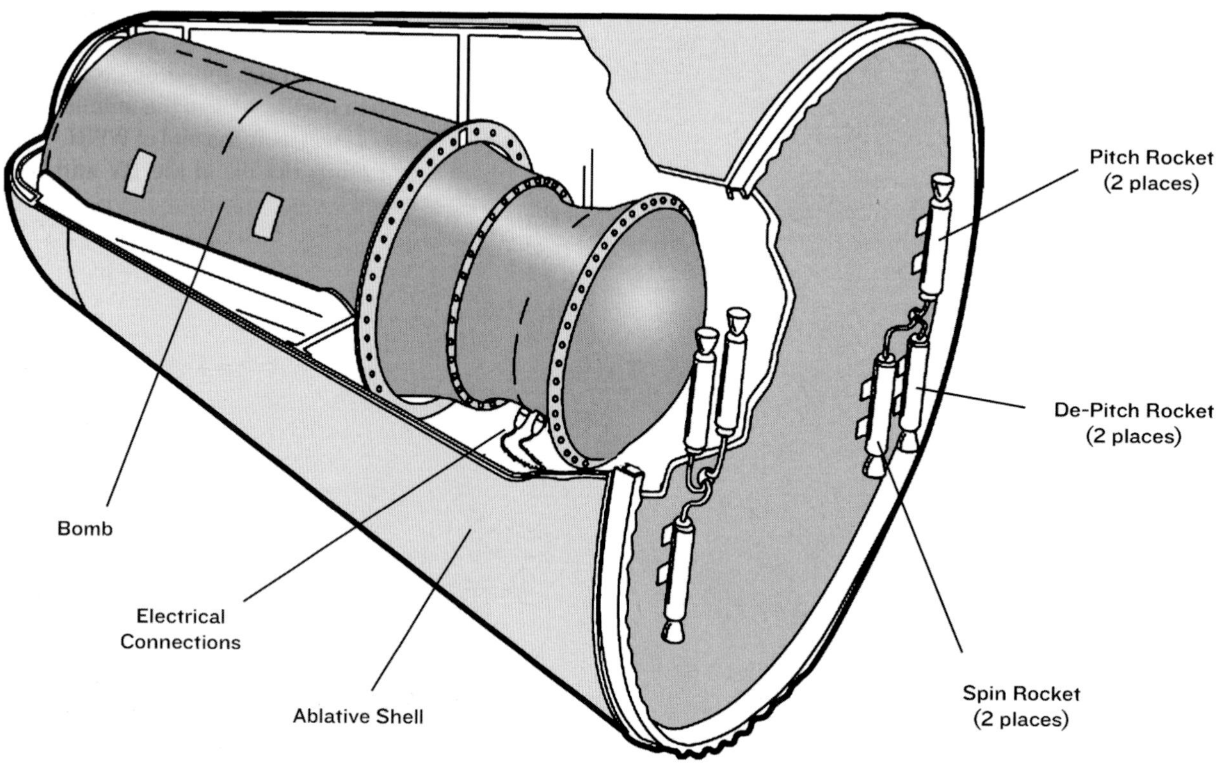

▶ This drawing of the presumed configuration of Mark VI and W-53 is based on information in the public domain and research by the author. Illustration by Larry Vance. Author's collection.

USAF PHOTO, TMM ARCHIVES.

▲ A maintenance team with a Mark VI RV mugs for the camera. This photo provides a rare glimpse of the RV's pitch and spin rockets, seen at right over the sergeant's left shoulder.

remain classified, but a number of facts are known. For example, the RV did not have a guidance system of its own, but it did have its own attitude control system. Thrusters mounted on the aft end of the RV nudged the vehicle into a nose-first orientation that prevented it from burning up in the atmosphere. Additional thrusters then give the RV a spin to stabilize its attitude.[27]

The detonation mode of the bomb was set before launch. It could be set to detonate as an air burst, or as a ground burst. A ground burst is defined as any detonation in which the fireball actually touches the ground.[28] Because a nine-megaton bomb would create a fireball about three miles (4.8 km) in diameter, a ground burst for Titan II could be a detonation as high as 7900 feet (2400 m).[29] That said, it is important to understand that there is an optimum burst height for a weapon of a particular yield that will maximize the destructive effects at a particular distance.[30] Declassified documents indicate the W-53 was set to detonate at an altitude of approximately 14,000 feet (4300 m) for an air burst and at actual impact on the ground for a ground burst.[31]

Perhaps the best and most accessible unclassified description of the bomb can be found in Chuck Hanson's remarkable book, *U.S. Nuclear Weapons*. This excerpt is reprinted with permission:

> The interior of the RV shell contained most of the arming and fuzing subsystems...magnetic inertial switches were located in the RV nose and a frangible impact–sensing printed circuit was installed throughout the shell.
>
> The arming and fuzing subsystem featured two- and four-element barometric switches, thermal batteries, timers, and a control box with optional ground- or air-burst selector. The arming and fuzing subsystem sensed both the presence and sequence of in-flight acceleration, deceleration, nose cone temperature, pressure, altitude, and course signals from the missile and the RV; if all inputs were correct in range and sequence then arming and fuzing signals were transmitted to the warhead.[32]

The RV would reenter the atmosphere traveling about 16,000 miles per hour (25,000 km/hr).[33] That works out to about 20 times the speed of sound, or just over four miles (six km) per second—considerably faster than a high-powered rifle bullet. At that speed, friction with the atmosphere would cause significant heating, and if the RV were not protected it would burn up just like a meteor. Protection came in the form of an ablative heat shield made from a plastic developed by General Electric. Ablative means that the heat shield was designed to burn away in layers, dissipating the heat as it did so. An even tougher version of the plastic was used on the nose, which would take the brunt of the heat.[34]

The W-53 was 3 feet (1 m) in diameter and 8.5 feet (2.6 m) in length. The bomb, together with its protective Mark VI reentry vehicle, weighed a total of 7575 pounds (3400 kg).[35]

Ballistic Flight and Accuracy

The entire flight of Titan II from launch to detonation at the target would take just 35 minutes, give or take, depending on the target. Of that time, only the first 5 ½ minutes were powered flight. After powered flight the RV entered the ballistic portion of its journey, speeding toward the target. Ballistic flight means that the RV was following the same kind of path as a bullet fired from a gun or as any thrown object.

When you fire a gun, the "powered" portion of the bullet's flight is very short, lasting only until the bullet clears the barrel. When you throw a ball, the powered portion of the ball's flight is that part before you let it go. After the bullet clears the barrel and after the ball leaves your hand, both are traveling a ballistic path. No matter if you aim the gun up, level, or anywhere in between, and no matter how hard, fast or far you throw the ball, the shape of the flight path is always the same, and is called a ballistic trajectory.[36]

So like an outfielder throwing the ball toward home plate, Titan II essentially threw the RV toward the target. After powered flight the RV continued to coast upward to approximately 800 statute miles (1300 km), following a ballistic path to the target.[37] This is why Titan II was called a ballistic missile.

The accuracy of a missile is spoken of in terms of circular error probable (CEP). CEP is defined as the radius of a circle into which bombs will fall in half of all attempts to hit the target. For example, half of all bombs delivered from ICBMs with a CEP of 1000 feet (300 m) will land within a circle 2000 feet (600 m) in diameter and centered on the target.[38] The requirement for Titan II called for a CEP of 1 nautical mile (1.15 statute miles, 1.85 km)—about the best one could hope for given the technology of the day. Operational tests confirmed that the missile performed to specifications.[39] Part of the reason Titan II carried such a high yield weapon was that its guidance system could only guarantee delivery to within the specified CEP. The bomb had to be large enough to destroy the target in the event it landed at, or just beyond, the CEP radius. Modern ballistic missiles carry weapons of much smaller yield because their guidance sets are accurate to within yards (meters) of the target.

Launching the Missile

Once a launch order had been received and authenticated, the MCCC and DMCCC calculated the time of launch. The MCCC then turned on the surface warning light and klaxon. Launch keys were removed from the EWO safe and installed in the key switches. The BMAT turned on circuit breaker 103, applying power to the missile's ordnance bus. Then the BVL code was entered in the butterfly valve coded-switch control. The code was called out by the MCCC and entered by the BMAT, who then read it back for verification. When the code was verified, the BMAT pressed the OPERATE INITIATE switch on the butterfly valve coded-switch control panel. If the code was correct

Discrete output signals issued by the MGC are:

1. GAN Flight Control System gain change
2. Spare
3. LOK Enable Thrust Chamber Pressure switch
4. SUS Sustainer Engine Cutoff
5. JET Jettison Decoys
6. VOF Vernier Engines OFF
7. ARM Prearm Warhead
8. REL Nose Cone Release
9. PIT-1 Pitch Rocket #1 ON
10. PIT-2 Pitch Rocket #2 ON
11. TON Translation Rockets ON
12. TOF Translation Rockets OFF

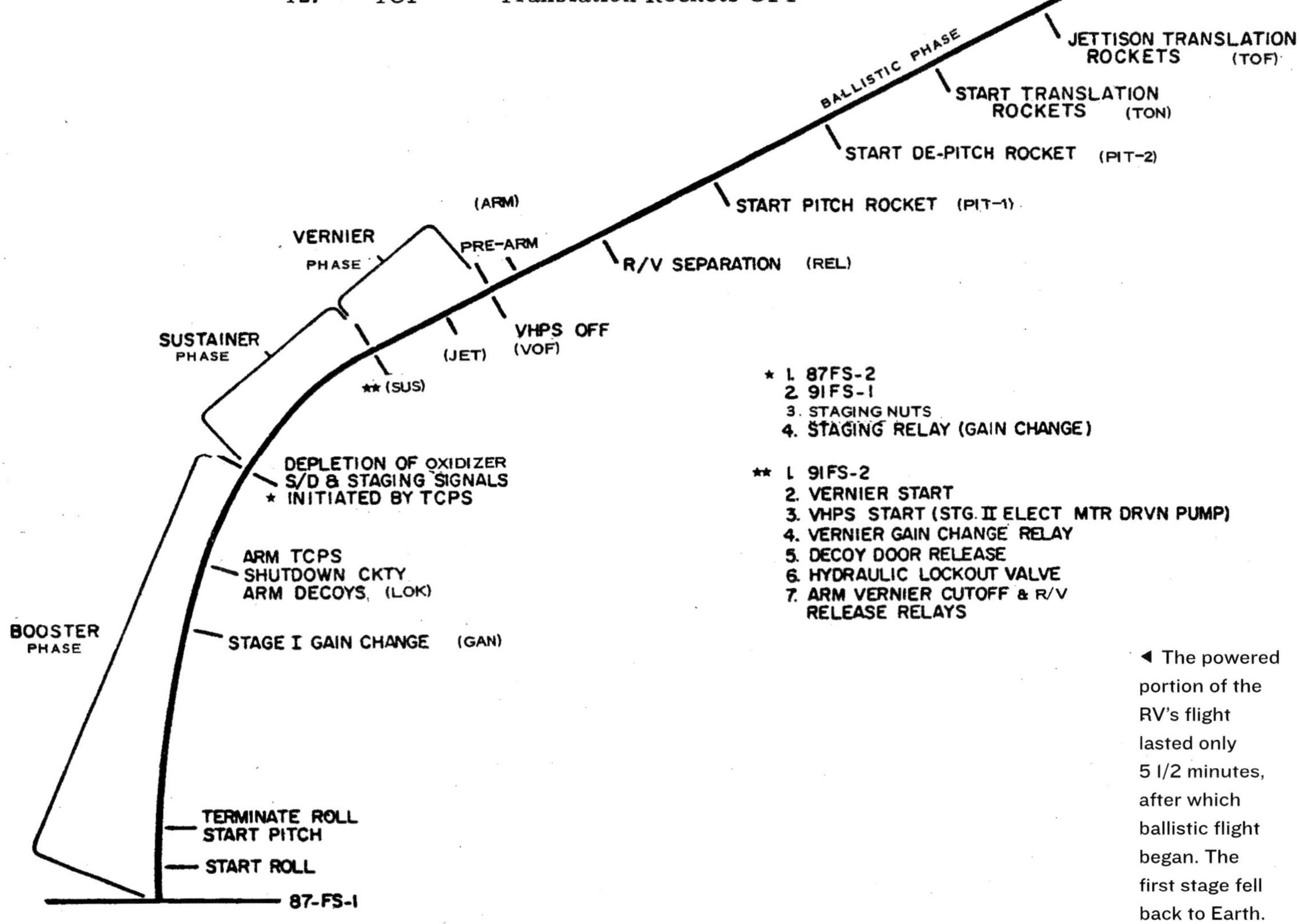

◀ The powered portion of the RV's flight lasted only 5 1/2 minutes, after which ballistic flight began. The first stage fell back to Earth. The second stage followed close behind the RV. Both reentered the atmosphere. After 1975 the translation rockets were not installed. No scale.

the OPERATE OK lamp lit up. This entire procedure could be done in less than three minutes.

To initiate the launch, two keys had to be turned. A key on the LCCFC was turned by the MCCC, while the other key, on the ALOC, was turned by the DMCCC. The key switches were spring-loaded to the off position and were physically far enough apart that one person could not reach them both. Moreover, the keys had to be turned within two seconds of each other and held in the turned position for five seconds to start the launch. This guaranteed that two people would be required to initiate the sequence. After the key-turn, everything else was automatic. There was nothing more the crew could do except to watch the process unfold. From key-turn to lift-off, the whole sequence took just 58 seconds.

Here's how it would have gone....

Launch Enable

Upon key turn, the LAUNCH ENABLE indicator illuminates, signaling that the launch sequence has begun. For all intents and purposes, this indicator might just as well have read WELCOME TO WWIII. At the same time, the inertial guidance system (IGS) MEMORY light turns on indicating that the onboard guidance computer is now operating independently and will no longer accept commands from the MGACG, such as changes in target or, under the old guidance system, AAS corrections to the gyros in the IMU.

Batteries Activated

There are batteries aboard the missile that supply power for the guidance computer and other electrical systems while the missile is in flight. Normally, these batteries are dry, like car batteries with no acid. A few seconds after LAUNCH ENABLE, the BATTERIES ACTIVATED light illuminates indicating that electrolyte is being forced into the batteries. About 28 seconds later, the batteries will be on-line.

During this interval a signal is sent to open the butterfly valves on the engines' fuel and oxidizer lines. Fuel and oxidizer flow into their respective turbopumps and purge any air from the propellant lines. To ensure that there are no air pockets or bubbles in the lines used to activate sequencing valves, a small quantity of fuel is actually dumped overboard into the launch duct.[40]

APS Power

When the batteries come on line, the APS POWER (accessory power supply) lamp confirms that the missile has switched to internal power.

At about this same time, the launch duct water deluge starts, and water begins pouring into the flame deflector. This causes an indicator on the FPCB to light DEFLECTOR NOT EMPTY.

Silo Soft

Almost simultaneously the SILO SOFT indicator lights up as the silo door begins to roll open. SILO CLOSURE NOT CLOSED lights to confirm the silo door is no longer in the fully closed position. If the air intake blast valve had not already been closed by the crew due to a higher state of readiness, the CC BLAST VALVES CLOSED indicator would light up as the control center blast valve closed with a dull thud. Seconds later, the tipsie intrusion alarm sounds because the motion of the door has disturbed the tipsie beam.

Guidance Go

Within a second or two of SILO SOFT, GUIDANCE GO lights up declaring that a check of the missile's flight control system has been completed successfully, that the missile's guidance system has taken full control of the mission, and that its internal clock has started, establishing "time zero." Time zero is the reference time from which all in-flight timed events will be calculated.[41] As the water deluge continues to drain from the 100,000 gallon tank, the SILO WATER SUPPLY LOW warning light comes on, as does the SILO SUMP WATER HIGH lamp, warning of water collecting in the flame deflector.[42] In a few more seconds, the silo door is fully open. Now, THRUST MOUNT SOFT indicates the thrust mount has locked into position, giving the missile a stable platform from which to launch.

Fire Engine

A second or two after THRUST MOUNT SOFT, a signal is sent to close the engine start switch, which lights FIRE ENGINE on the LCCFC. Instantly the stage-one engine start-cartridges ignite, spinning up the propellant turbopumps. In the blink of an eye the engine roars into life, transforming the launch duct into a thundering inferno. The fire and heat are so intense that the steel reinforced concrete of the flame defector begins to burn away. Fire sensors in the launch duct send a signal to sound the LCC fire klaxon.

Several indicator lights warn of fire and high temperature in the launch duct. About two seconds after starting, the engine has developed enough thrust for lift off and Titan II begins to strain against its hold-down bolts. When pressure sensors confirm a predetermined thrust level, hold-down timers wait about two seconds to make sure that engine thrust is both stable and symmetrical. If all is well when the timers run out, a signal is sent to explode the hold-down bolts.

Lift Off

Gravity may be the law, but it is no match for the hypergolic fury of Titan II's Aerojet engine. When the hold-down bolts explode, Titan II virtually leaps away from the thrust mount. An instant later the LIFT-OFF and UMBILICALS NOT CONNECTED indicators illuminate as lanyards attached to the umbilical cables on the missile stretch tight and pull the cables away from their sockets. As the missile climbs out, hoses, wires, cables, sensors and acoustical tiles in the launch duct disintegrate in the firestorm. Less than five seconds after the hold-downs let go, all 165 tons (150 metric tons) of Titan II is clear of the silo and on its way. Sixty seconds later the missile is but a star-like speck in the sky.

Thirty-five minutes and 6300 statute miles (10,000 km) distant, Titan II's target awaits Armageddon.

After the launch, the crew starts the post-launch checklist—and hopes for the best.

1 *Weapon System Familiarization*, Course OZR1821F/3121F, Technical Training, Shepard AFB, January 1965, 58

2 Stumpf, 49

3 00-ALCP 144-4, table A6-1

4 DELETED

5 http://www.boeing.com/commercial/747family/pf/pf_400f_prod.html

6 This time is based on video of Titan II launches.

7 TO.21M-LGM25C-1, 1-134, 1-135

8 Stumpf, 178

9 Ibid., 67-68

10 TO.21M-LGM25C-1, 1-72

11 *Weapon System Familiarization*, 175

12 TO.21M-LGM25C-1, 1-124

13 Stumpf, 49-50

14 Ibid., 178

15 Primacord is a trademarked name for a rope-like linear explosive made by Dyno Nobel Inc. See http://en.wikipedia.org/wiki/Primacord

16 On N-10's airframe these are labeled "retro rockets." On missiles examined at AMARG they are labeled "pitch rockets."

17 The Martin Company, *Detailed Design Specifications for Model SM-68B*, 1-September-1961, updated through Change Notice 66, 24-January-1969, 25-26

18 The ABM treaty permitted each side to defend only its national capital and one ICBM field. The U.S. never pursued an ABM system to protect Washington, but a system built to protect Minuteman missiles around Minot, North Dakota was already 85 percent complete when the treaty was signed. This system was operated for only one year before being recognized as useless against the hundreds of Soviet warheads that could be launched against it. It was abandoned in 1976. See John Lonnquest and David Winkler, *To Defend and Deter: The Legacy of the Cold War Missile Program*, Department of Defense Legacy Resource Management Program, 1996, 111-117. See also Gerald Clarfield and William Wiecek, *Nuclear America: Military and Civilian Nuclear Power in the United States*, 1940-1980, 1984, 289-343.

19 TO.21M-LGM25C-1, 1-135

20 Ibid., 1-135

21 Stumpf, 64-65

22 Ibid., 66-67

23 http://www.old-computers.com/museum/computer.asp?c=252

24 Delco Electronics Corporation, *System Technical Description*, S86-147, July 1986

25 Stumpf, 62

26 Hansen, 199

27 Air Training Command, OTS31000-14-I-1, *Supervisors and Planners Course WS107A-2 (SM-68B)*, 22A-22B, November 1961

28 Glasstone, 10-12

29 Nuclear Bomb Effects Computer, 1962 revision, based on data from "The Effects of Nuclear Weapons"

30 Glasstone, 130-132

31 Strategic Air Command 1st Aerospace Division, *Operational Test Final Report for the LGM-25C Weapon System (Titan II)*, 31 May 1966

32 Hansen, 199

33 Stumpf, 178

34 Ibid., 56-63

35 *Detailed Design Specifications for Model SM-68B*, 16

36 http://en.wikipedia.org/wiki/Trajectory#Physics_of_trajectories

37 TRW-STL Final Report, 15

38 Kosta Tsipis, *Arsenal: Understanding Weapons in the Nuclear Age*, (Simon and Schuster 1983), 139-144

39 Strategic Air Command 1st Aerospace Division, *Operational Test Final Report for the LGM-25C Weapon System (Titan II)*, 31 May 1966

40 Aerojet Liquid Fuel Rocket Company, *Titan II Propulsion Subsystem*, 1981, 3.3-3.4

41 TO.21M-LGM25C-1, 1-137

42 As observed in Air Force Training film "How Would You Feel?" TMM reference video #43.

CHAPTER 8

AFTER THE LAUNCH

Two of the most frequently asked questions at the Titan Missile Museum are "How long could the crew survive underground?" and "How close could a bomb fall without destroying the complex?" The standard reply to the first question is that the crew had 30 days of food, water, air, and fuel for the generator. The second question is more complex.

These outwardly simple questions are, in fact, exceedingly complicated. There are so many variables in the equations of crew and complex survivability that no definitive answers are possible. Still, these are important questions and ones that deserve closer examination.

After a launch the crew had standing orders to wait for further instructions. These instructions would almost certainly come by telephone or radio, although it is conceivable that someone could drive to the complex and call from the phone at the main gate, assuming the gate phone had not been destroyed, that working vehicles still existed, and that roads were passable.

There was a "recovery and reconstitution plan" but it had nothing to do with crew survival. It concentrated on how crews who had launched their missiles could help crews who had not. The plan offered some suggestions on how this might be done, but conceded that crews would have to do a lot of improvisation.

Still, if after thirty days the crew had had no communications from the wing command post or other authority, they were essentially on their own, and had to decide for themselves what to do next.

An important point to bear in mind in this discussion is that if the crew had been asked to launch it meant that WWIII had begun. This, in turn, meant that nuclear weapons were exploding on U.S. soil, perhaps in large numbers.

Fallout

The primary reason the crew would want to stay underground after a nearby nuclear detonation is radioactive fallout. The amount of fallout that might cover the complex would be determined by where and how the bomb detonated, and the direction of the wind. If the detonation was nearby and the silo was downwind, fallout could be heavy. If the detonation was more distant, and/or if the silo was not downwind, fallout could be much lighter. While air bursts create less fallout than ground bursts, it seems reasonable to assume that the launch sites would have been targeted with ground bursts.

Complicating the crew's decision on whether or not to leave the protection of the complex is that they had no way of assessing conditions on the surface without actually going up to have a look. Even then, they did not have a Geiger counter or even a simple dosimeter to help them determine the level of radioactivity. Human beings cannot directly sense radioactivity. It could be a bright and sunny day and still be lethally radioactive. Why crews were not issued simple radiation detecting equipment of the kind found in every civilian fallout shelter of the day is not known. This kind of equipment was reliable, durable, readily available, and so cheap as to be essentially free. It would be tempting to suggest that the Air Force regarded the crews as expendable, but this would be overly simplistic. With access to radiological equipment, not only could crews have assessed the risks to themselves, they could also have provided information essential in reconstitution efforts—regarded as a high priority as evidenced by the existence of a detailed plan —and could have provided a wide area assessment of fallout that would have been instrumental in troop mobility, rescue operations and damage assessment.

Radiation levels from fallout decrease rapidly with time. As a rule of thumb, the level falls by a factor of 10 with every seven-fold increase in time.[1] For example, seven hours after the fallout had stopped falling radiation would have decreased to one-tenth its original value. After 49 hours (7x7) the level would be just one percent or the original, and after 343 hours (7x7x7), or about two weeks, the level would be only a thousandth of its original intensity. Although this may sound reassuring, the initial level of radioactivity may well be high enough to be unsafe even when reduced by a factor of 1000. Furthermore, even when radiation had fallen to non-lethal levels, some contaminated areas would not be considered safe by peacetime standards for a decade or longer. Exposure to radioactivity is not unlike exposure to the sun. The effects are cumulative—the longer you lie on the beach, the redder you become, even if you only spend a short time on the beach each day. Day after day, it adds up. Even at low levels of radioactivity, damage to your body accumulates the longer you are exposed.

The Essentials

There are four commodities that would have limited the crew's length of stay in the LCC. These are food, water, air, and to a lesser extent, diesel fuel.

The crew had a 30-day supply of canned food and water, in addition to 124 gallons (470 l) of potable water in a tank on LCC level 1. Still, if the crew intended to stay for 30 days, some water rationing would be essential. With an adequate supply of canned drinking water, the 124-gallon water tank could be reserved for basic sanitation, and the largest single use would be for the toilet. Assuming three gallons (11 l) per flush, the four-person crew, collectively, would be limited to one flush per day, if they used the water for no other purposes. A resourceful crew might decide to remove the cleanout hatches from the sewage ejection pump collection tank and use it as a kind of makeshift outhouse.

With an unknown amount of radioactive fallout topside, operating the fresh air intake fan for the LCC would be risky at best (assuming electric power was available to run it), even with an air filter in the ducting. Therefore, the crew would be limited to the air contained in the LCC. The average person breathes about a half a liter of air with each breath taken. At a rate of 15 breaths per minute the average person would inhale nearly 11,000 liters of air per day.[2] The four-person crew, then, would take in a combined total of 44,000 liters of air per day. The control center contains about 750,000 liters of air, so that in 17 days all the air in the control center would have been breathed once. In theory, at least, that should leave enough oxygen to breath all the air again. A more sophisticated calculation would account not only for a *decreasing* amount of oxygen in the air, but also for an *increasing* amount of carbon dioxide, since carbon dioxide toxicity would occur well before the oxygen ran out. In any case, a month in the LCC seems doable for four people, if only just.

It should be noted that a nuclear attack might continue for a number of days, and while the "air clock" would start ticking from the instant the LCC blast valve closed (which may have happened anywhere from several hours to a few days before the actual launch), the "radiation hazard clock" would not start until the attack had ceased. The difference in time could be significant in the crew's survival.

Of course, all of these calculations assume only four people inside the LCC. It is conceivable that in addition to the crew, the pad chief, security team members, and/or other crew members might be present. The addition of extra people would change the equation considerably. On the other hand, if one or more of the crew were somehow killed during an attack, the equation changes as well.

Finally, there is the matter of fuel for the generator. As discussed earlier, post-attack use of the generator would be problematic. The crew would have one of three quantities of fuel available, depending on the extent of damage to the complex: 10,500 gallons (40,000 l) if the near-surface and underground tanks both survived *and* the crew had access to the switch on level 5 *and* the transfer pump timer in the access portal was not damaged; 2500 gallons (9500 l) if the near-surface tank was damaged, or the crew could not get into the silo, or if the access portal had collapsed; or 55 gallons (208 l) if both tanks were unavailable (refer to the discussion on pages 104-105). In terms of generator run time these figures translate roughly to 420 hours, 100 hours, and 2 hours, respectively, assuming the generator itself was not damaged and could be started. For comparison, there are 720 hours in 30 days, so if all the fuel were available the generator could be run for about 14 hours per day. With only 2500 gallons available, use of the generator would be limited to about 3 hours per day. Bear in mind that the generator would have to be operated at intervals short enough to keep it from cooling down so much that it could not be started again, and when limited to 3 hours total run time per day, that could be challenging. Operation of the generator also depended on a reliable supply of cooling water. This would not be a "given" in a post-attack senario.

Effects on Crew and Complex

The effects of a nuclear detonation on the crew and complex are far more difficult to quantify. While a great deal of hard data on nuclear weapons effects exists as the result of hundreds of U.S test detonations at the Nevada Test Site and elsewhere, the launch complexes themselves were never directly subjected to a nuclear test, and structural effects data from weapons of the size likely to have been delivered upon the sites may be regarded as somewhat sketchy.

Still, the available data provide some insight into the consequences of a nuclear detonation in the vicinity of the complex. First, it is important to understand that no known structure is capable of withstanding a direct hit. In the case of a detonation of almost any yield on or just above a launch site, the entire complex, all those within, and the missile would be vaporized, or at minimum, smashed into oblivion. Detonations further afield would inflict varying degrees of damage.

How Close is Too Close?

To answer this question two numbers must be known. First, the maximum overpressure the complex can withstand, and second, the distance from the point of detonation at which that overpressure would not be exceeded.

Yield of Weapon	Approximate Radius of 300 psi Gradient	Approximate Crater Diameter
20 MT	5800 ft (1800 m)	3400 ft (1000 m)
10 MT	4700 ft (1400 m)	2800 ft (850 m)
1 MT	2200 ft (660 m)	1300 ft (400 m)
500 KT	1600 ft (500 m)	1000 ft (300 m)

As discussed earlier, Titan II launch complexes were hardened to withstand an overpressure of 300 psi (21 kg/cm^2) applied to the silo door and the surface of the ground above the site. At close range, there isn't much difference between ground bursts and air bursts, at least in terms of overpressure gradients. The chart above displays the distance to which a 300 psi overpressure would extend (for air or ground burst), and the approximate size of the crater formed by a ground burst.[3]

Any detonation closer than these distances would be considered a direct hit. The actual distance would depend on a number of factors including, for example, local terrain, but from the chart it becomes clear that a 1-megaton bomb could "miss" the complex by four tenths of a mile and still inflict sufficient damage to put it out of operation. There are several ways in which the mission could be scuttled by excessive overpressure. These include induced ground shock that could overwhelm the missile suspension system, shock-induced failure of the silo door, shock-induced failure of electrical or other critical systems, physical damage to or destruction of the LCC or silo structures, and large amounts of ejecta covering the silo door, to name only a few. If the missile were damaged and exploded in-silo due to rupture of the propellant tanks, for example, the crew would be unaffected, assuming the blast doors were closed.

Because the 300 psi (21 kg/cm^2) overpressure range greatly exceeds the size of the crater formed from a ground burst, worry of finding yourself inside the crater becomes a moot point, especially given the relatively shallow depth of the complex.[4] And compared to overpressure, the threat from direct ground shock is actually quite small only a short distance beyond the crater rim. Induced shock from a pressure wave sweeping over the surface represents a substantially greater hazard.[5]

A shock-isolated control center notwithstanding, as long as crew members were not thrown in such a way as to incur significant injury, and as long as they were not struck by concrete falling from the top of the LCC dome, for example, they may escape the attack with relatively minor damage.

In addition to overpressure and shock, a nuclear detonation also produces an intense burst of neutrons and gamma rays, both of which are very dangerous and highly penetrating. The complex provided substantial shielding from both. The 18-inch (46 cm) -thick domed roof of the LCC and the nine feet (2.7 m) of earth above it provided shielding of a factor of more than 10,000. In a detonation close enough to provide a lethal dose of neutrons or gamma rays in the LCC, damage from overpressure would probably provide a more immediate demise.[6]

Final Thoughts

In a nuclear exchange the crew and complex would be subjected to forces and effects on a scale that is difficult to imagine, and to events that are impossible to predict. The combined effects of EMP, shock waves, neutrons, gamma rays and fallout taken together with the physical and psychological trauma of a nuclear war would test human endurance in a way that no one can fully understand. So what is the answer? Could the crew survive a nuclear exchange? Outside of a direct hit, yes, probably, depending on the breaks. But in the final analysis that is not the central question. If the crew survived the initial events and managed after 30 days to stumble out of a dark and foul smelling complex into the light of day, they would find themselves face to face with a post-nuclear wasteland in which hundreds of millions of people lay dead or dying and in which civilization had been effectively crushed. They would find themselves in a world so profoundly changed that, for all intents and purposes, they would be living on a different planet. The world they knew before they turned the keys would be gone. Forever. Yes, the crew might survive, but the more pertinent questions would be: where would they go, how would they get there, and finally, what would become of them?

1 Glasstone, 418-424

2 http://en.wikipedia.org/wiki/Lung_volumes

3 Glasstone, 130-148

4 Ibid, 289-290

5 Ibid, 280-281

6 Ibid, 10, 386-394

CHAPTER 9

NUCLEAR WEAPONS 101

No discussion of Titan II would be complete without some basic understanding of the nuclear weapon it carried.

Nuclear weapons are thousands or even millions of times more powerful than the largest bombs made with conventional high explosives. The horrifying efficiency of nuclear weapons made it possible to destroy entire cities with a single blow and changed the basic thinking about how wars might be fought. Unlike the years-long conflicts of World Wars I and II, World War III could be over in as little as 90 minutes.

How They Work

Nuclear weapons come in two basic designs: fission devices, casually referred to as atomic bombs or A-bombs; and fusion devices, often called thermonuclear bombs, hydrogen bombs or H-bombs.

Most people are familiar with Albert Einstein's famous equation, $E=MC^2$. This equation says that energy (E) is equal to mass (M) times the speed of light (C) multiplied by itself. The only number you have to know here is the speed of light: 186,000 miles (300,000,000 m) per second. You don't have to know much about math to figure out that when you multiply anything by the speed of light squared you get a huge number. What Einstein's equation means is that a very small amount of matter contains a very large amount of energy.

Even individual atoms contain a substantial amount of energy. When atoms split (a process called fission) some of that energy escapes. Atoms of very heavy elements like uranium and plutonium are comparatively easy to split, and these became the fuel of choice for atomic bombs. When atoms of these heavy elements undergo fission, the combined weight of the two resulting lighter atoms is a little less that the weight of the original. The "lost" weight is converted to energy.[1,2] And from the fissioning of zillions of atoms all at once you can get a very big bang. That, in a nutshell, is how atomic bombs work.

In addition to releasing energy, when an atom of uranium or plutonium is split, it also ejects two or three particles called neutrons. If these neutrons collide with other atoms of uranium or plutonium, those atoms can split as well, releasing more energy and more neutrons to split more atoms. These neutrons are the key to nuclear power because they make possible a self-sustaining chain reaction. Every split atom has the potential to split at least two more atoms, and so the number of atoms split increases exponentially: 2, 4, 8, 16, 32, 64 and so on. The chain reaction happens very quickly (in an atomic bomb the entire process

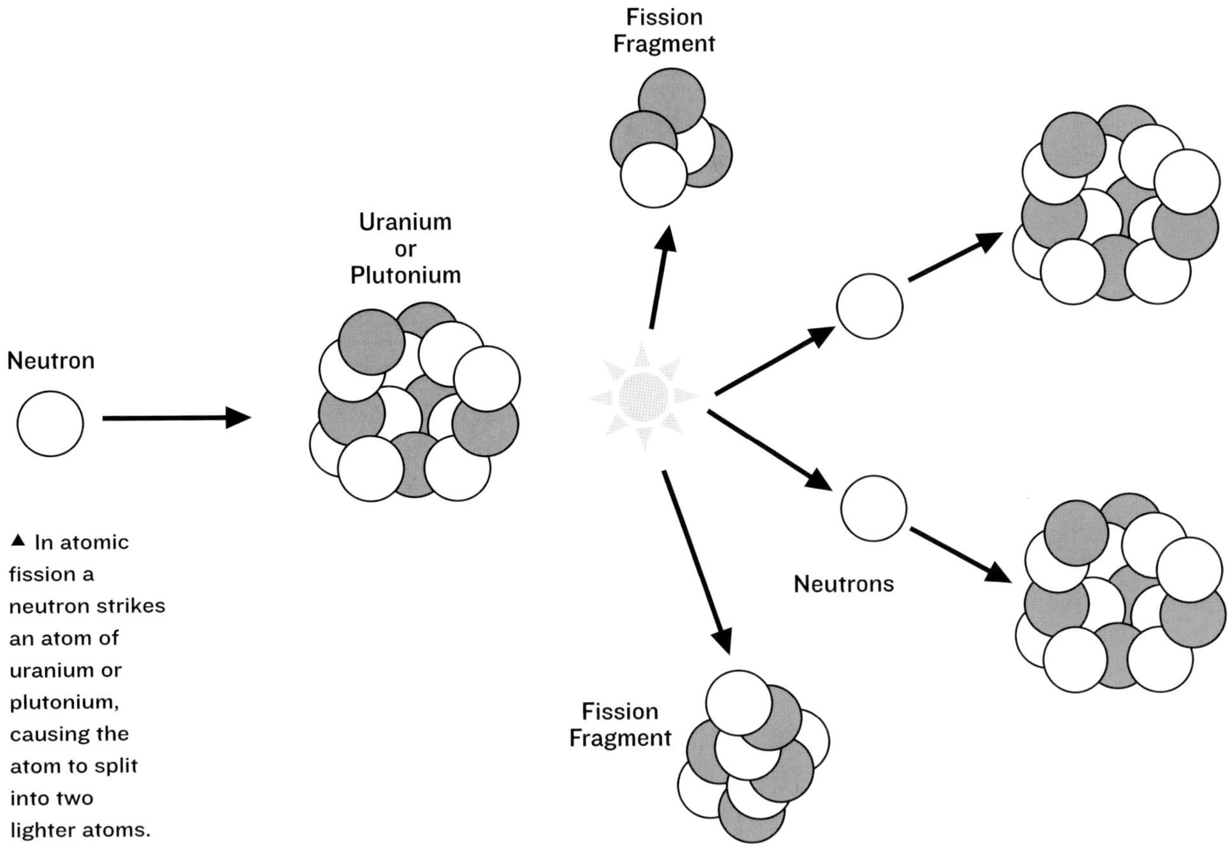

▲ In atomic fission a neutron strikes an atom of uranium or plutonium, causing the atom to split into two lighter atoms. This releases energy and two or three neutrons that strike other atoms and continue the process.

is over in about one millionth of a second) releasing immense quantities of energy. It all sounds simple enough, but it took dozens of the brightest minds in the world more than two years to work out all the details.

Creating a chain reaction requires a minimum quantity of fissionable material, referred to as a critical mass. Any amount of uranium or plutonium of less than this minimum mass will not support a chain reaction and is said to be sub-critical. A critical mass is one that will sustain a steady-state chain reaction—one that provides a constant energy output. What is needed for a nuclear weapon is a super-critical mass—one that will provide a "runaway" chain reaction with an ever-increasing output of energy. The idea in an atomic bomb is to create a super-critical mass from one that is sub-critical. The trick is to create the mass quickly enough so that the chain reaction will be as large as possible before the mass blows itself apart. Two techniques were devised to accomplish this.

The first was a simple gun-type device in which a sub-critical slug of uranium is fired into a sub-critical uranium target. The slug is propelled toward the target with a chemical explosive charge, just as a bullet is propelled from its casing. When the slug hits the target the super-critical mass is established. Boom. This technique works well with uranium, but is too slow to use with plutonium. Plutonium fissions so easily the chain reaction would begin long before the two pieces met. The result would be a fizzle.

In a bomb using plutonium, a sub-critical sphere of plutonium is surrounded by a sphere of chemical explosives. This is often called an implosion device. When detonated, the shockwave of the explosion compresses the

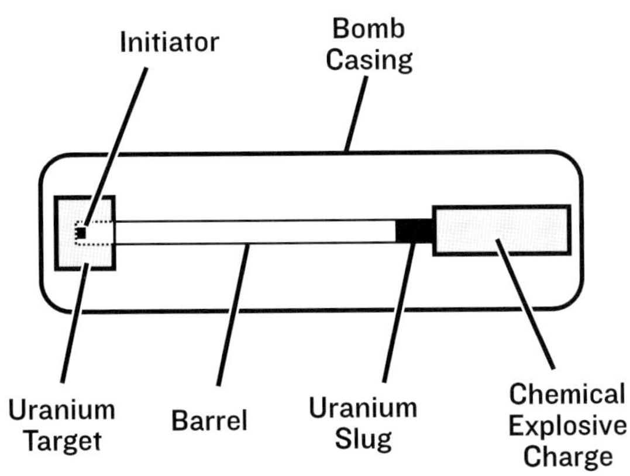

▶ Simplified schematic of a gun-type atomic bomb. The slug is fired toward the target at bullet-like speed. When the two meet, a super-critical mass is formed.

AUTHOR'S ILLUSTRATIONS.

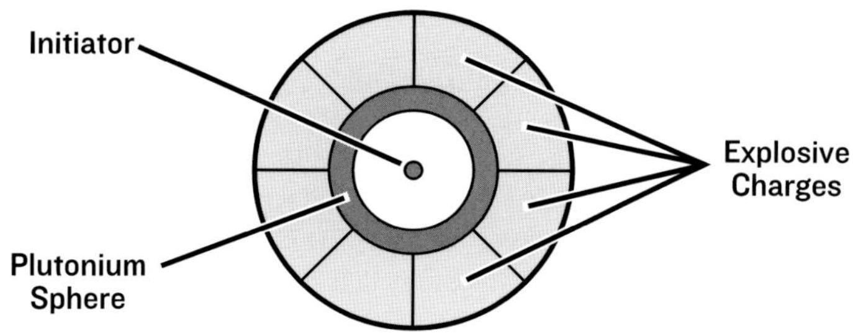

▲ Simplified schematic of an implosion device. The shockwaves of the chemical explosives compress the plutonium sphere to super-criticality. Author's illustration.

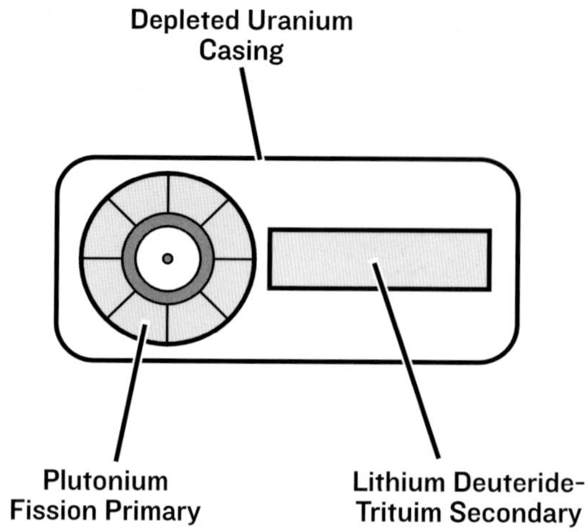

◀ Highly simplified schematic of a fusion bomb. The fission primary explodes and compresses the fusion secondary. Excess neutrons created by fusion are used to create fission in the bomb casing.

sphere of plutonium until it becomes supercritical.

In both uranium and plutonium devices an initiator is used to spray additional neutrons into the mass in order to cheat the chain reaction ahead by many generations, thus boosting the yield.

Fusion weapons aren't much more complicated, but instead of splitting atoms, they force two atoms of light elements (like hydrogen) together so they combine, or fuse, into a single atom of something a little heavier.

Fusing two atoms together is not a simple task. It requires temperatures as high as at the core of the sun— 27,000,000 °F (15,000,000 °C).[3] To achieve this, a small fission bomb, called the primary, is used as a trigger to start the fusion process in the secondary. In fusion the weight

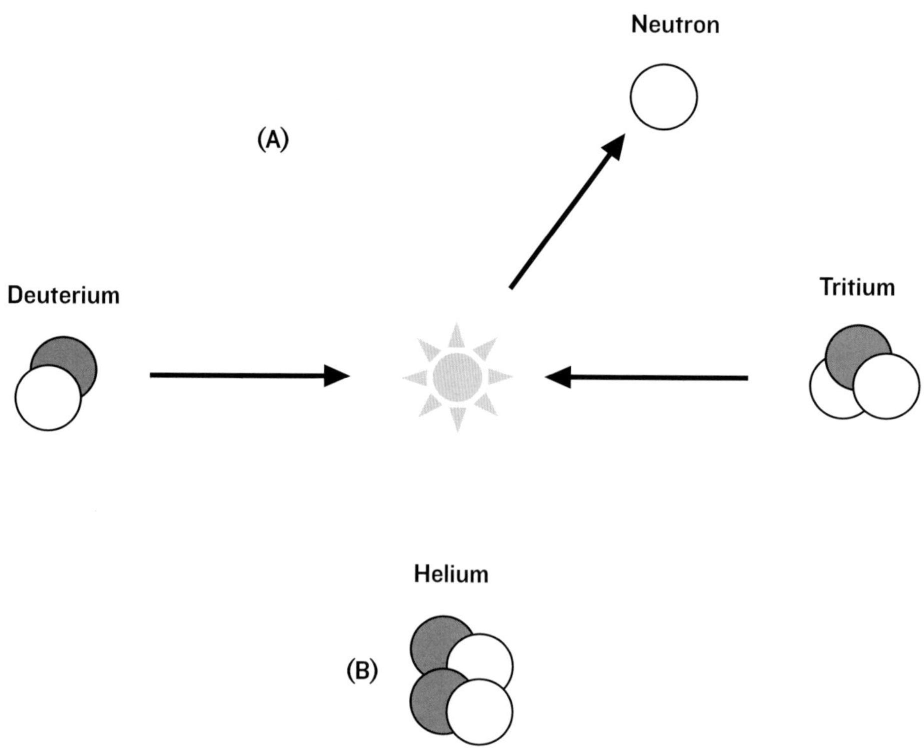

▶ (A) Deuterium and tritium are forced together under extreme heat, creating a helium nucleus and releasing energy. Since only two neutrons are needed in a helium atom (B), the excess neutron is ejected from the process.

AUTHOR'S ILLUSTRATIONS.

NASA ARCHIVES.

▲ The Mike device proved the concept of fusion, and made possible high-yield lightweight nuclear bombs. The vertical tank at left was filled with liquid deuterium and was the part that actually exploded. The many horizontal pipes transmitted light from the explosion to scientific instruments.

of the resulting heavier atom is a little less than the total weight of the two original atoms. Again, the missing mass is converted to energy, but in the fusion process a great deal more energy is released than in the fission process. In fact, pound for pound, fusion creates about three times more energy than fission.[4] To create a little more bang for the buck, fusion bombs are often contained in a case made of depleted uranium. The temperature and neutrons generated by fusion cause fission of the casing, adding to the total explosive force.

Strictly speaking, hydrogen bombs do not use hydrogen. They use isotopes of hydrogen, specifically deuterium (a hydrogen atom with one neutron) combined with lithium to form lithium deuteride, and tritium (a hydrogen atom with two neutrons) [5,6,7]

Yield

The explosive power of nuclear weapons, whether fusion or fission, is called the yield. Yield is measured in units called kilotons or megatons. These units refer to an equivalent explosive force in TNT, a chemical high explosive. A one-kiloton yield therefore equals the force of one thousand tons of TNT, while a megaton equals one million tons of TNT. Early fission bombs had yields of 15 to 20 kilotons. The yield of fusion weapons can be greater than a megaton, but in modern weapons systems, including currently deployed missiles like Minuteman and Trident, the yield is generally between 150 and 350 kilotons.

Kilotons and megatons are convenient measures for describing the yield of nuclear weapons but they unintentionally disguise their terrifying destructive power. How big is a megaton? What does that really mean? To put yield into perspective consider the length of a freight train required to haul a single megaton (one million tons) of TNT. Using standard 100 ton (90 metric ton) freight cars, and assuming that 100 tons of TNT would actually fit in a 100 ton freight car, then a freight train hauling one megaton of TNT would be on the order of 125 miles (200 km) long.[8]

The bomb carried on Titan II was a fusion weapon with a yield of nine megatons, or if you prefer, the equivalent explosive power of a TNT-packed freight train about 1100 miles (1800 km) long. That is enough power to lay waste to an area of more than 900 square miles (2300 km^2.)[9]

The Effects of Nuclear Weapons

The type and severity of destruction wrought by a nuclear weapon depends on many factors: the yield of the bomb, whether the explosion is a ground burst or an air burst, the altitude of the detonation, proximity to the target, the

DEPARTMENT OF ENERGY PHOTO.

◄ The fireball is actually air that has been heated to incandescence.

type of terrain on or over which the explosion occurs, the type of soil, the kind of target and local weather conditions, to name a few. Still, it is possible to generalize.

Nuclear weapons unleash their fearsome power in three ways: thermal radiation (heat), blast, and radiation over a broad spectrum from light to gamma rays and X-rays. While the proportion of each of these depends to a certain extent on the type and yield of the weapon, in general about 50 percent of the energy is released as blast, 35 percent as heat, and 15 percent as radiation. This last category can be broken down into two types: prompt radiation, including gamma rays and neutrons which are emitted in a quick burst at the time of detonation; and residual radiation that lingers for days, months and even years. Residual radiation takes the form of radioactive fallout created when vaporized bomb materials and their by-products combine with dust and dirt swept up in the mushroom cloud.[10,11]

In a hydrogen bomb the fissioning of the

DEPARTMENT OF ENERGY PHOTO.

▲ The classic mushroom cloud of a nuclear explosion is caused when air rushes in toward the low pressure zone created at the center of the detonation. Because hot air is less dense than cool air, the fireball is buoyant, and rises rapidly. The streamers seen at left are made by small rockets fired just before the bomb is exploded, and are used to observe the shockwave at various altitudes. All explosions, even very small ones from fireworks, create mushroom clouds.

uranium or plutonium primary and subsequent fusion of the tritium-deuterium components liberates vast amounts of energy in less than a millionth of a second. This energy vaporizes all the material that once composed the bomb, transforming it into a mass of super-hot gas under tremendous pressure—millions of pounds per square inch. These hot gasses emit a burst of high energy X-rays that are quickly absorbed by the air surrounding the bomb. The energy of these X-rays is transferred to the air, heating it to incandescence, resulting in the rapidly expanding fireball seen in nuclear detonations.[12]

The creation of the fireball has two significant effects. First, the heat from the fireball radiates outward at very near the speed of light as a thermal pulse. The closer one is to the point of detonation, the greater the intensity of the pulse. There are many variables that play into the thermal effects of a nuclear weapon so it is difficult to say with any precision exactly what would happen at some particular distance. That said, it is likely that a nine-megaton weapon would start significant fires as far away as 20 miles (30 km),[13] inflict third degree burns on exposed skin at nearly 25 miles (40 km) and first degree burns at 30 miles (50 km) or more. Additionally, the sheer brilliance of a

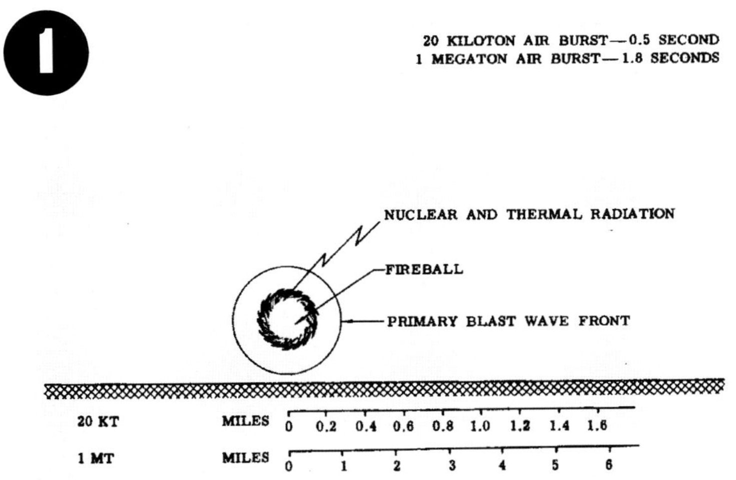

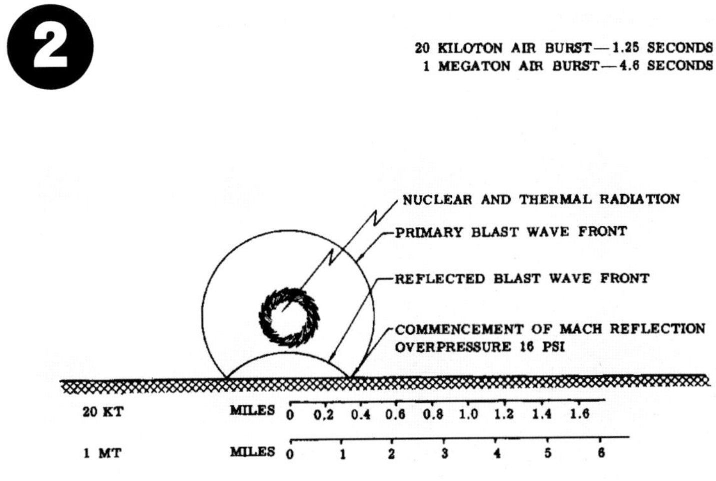

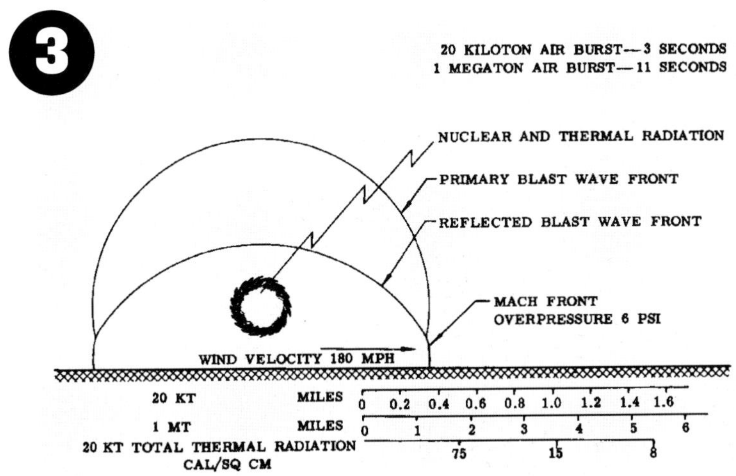

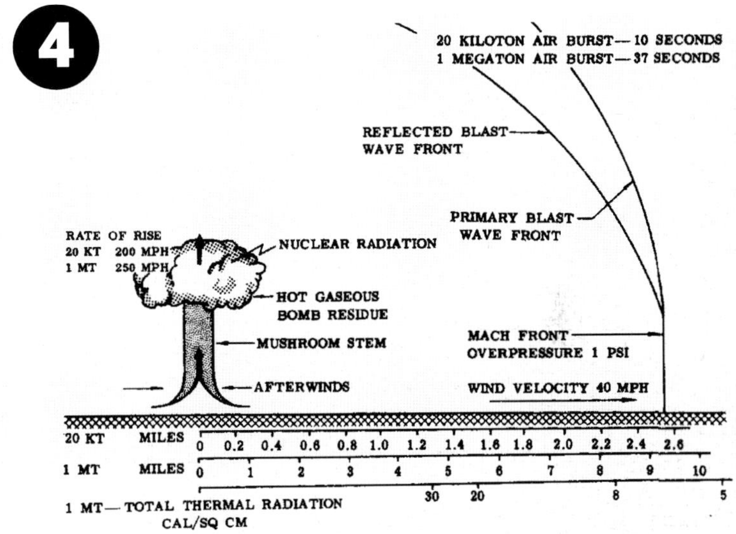

▲ Diagrams illustrating the formation of the fireball, blast wave, and mushroom cloud. Note time after detonation and distance scales.

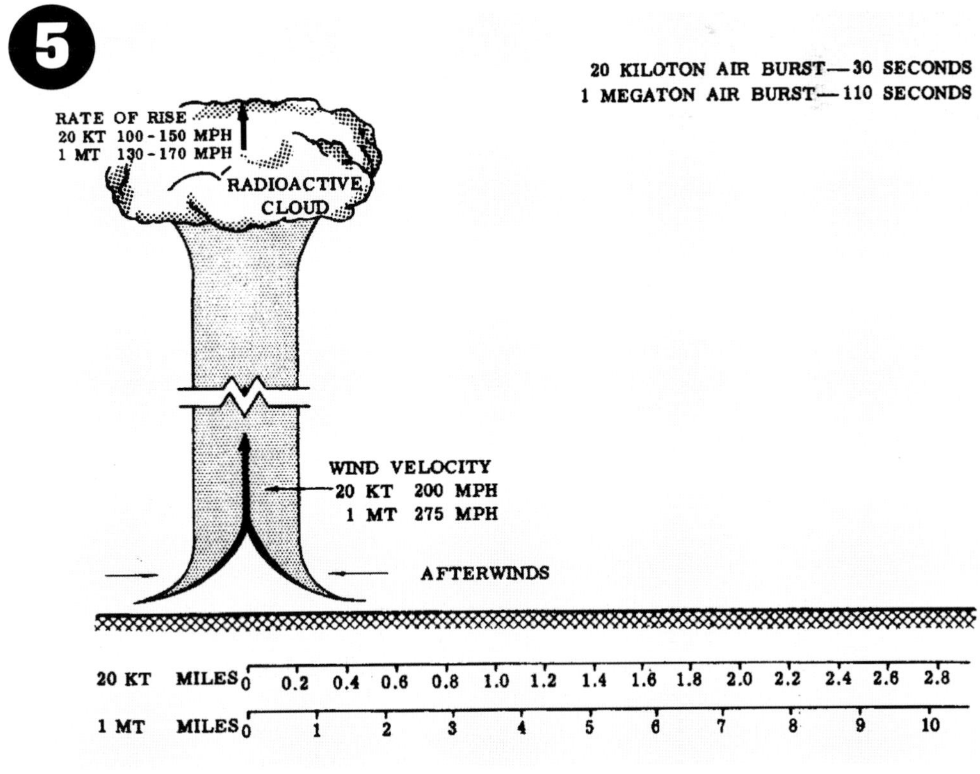

FROM "THE EFFECTS OF NUCLEAR WEAPONS."

nine-megaton fireball would be many times brighter than the sun at noon even at distances well beyond 100 miles.[14,15] Anyone unfortunate enough to be looking in the general direction of the blast when it occurred would almost certainly suffer temporary or permanent blindness as a result.[16]

The second significant effect of the fireball is *blast*. Blast from nuclear weapons is caused by the rapid expansion of super-heated air. As the air expands outward from the point of detonation it can be thought of as a moving wall of highly compressed air. Near the point of detonation this moving wall of air would be traveling at supersonic speeds.[17] The blast wave is often referred to as a shock wave because the effect on objects it encounters is that of a sudden impact. Depending on the height of the burst, blast waves approaching the ground may encounter waves reflecting up from the ground and combine to form a "mach front" with an overpressure nearly twice that of the individual waves and sufficient to flatten virtually everything in its path.[18] In addition, high winds, the velocity of which depends on the pressure of the blast wave at any particular point along its path, follow the passage of the blast wave or mach front.

Blast waves generated by any explosion exert a powerful force on objects they encounter. This force is referred to as overpressure. Overpressure is the amount of pressure on the surface of an object above normal atmospheric pressure. Overpressure is measured in pounds per square inch (psi) and overpressures of just three to five psi (0.21 to 0.35 kg/cm^2) will inflict severe damage on wood-frame houses. Overpressures from five to 10 psi (0.35 to 0.70 kg/cm^2) would collapse most non-reinforced concrete or brick residential-type structures even if their walls were 12 inches (30 cm) thick. Not very many structures of any kind could survive overpressures above 10 psi.

As a point of reference, the blast valves that protect the crew in the LCC are designed to close automatically with two psi of overpressure. While two psi of overpressure may not sound like very much, consider the following description from a PBS documentary on the subject:

> Any single-family residences that have not been completely destroyed [by the two psi shockwave] are heavily damaged. The windows of office buildings have been blown away, as have some of their walls. The contents of these buildings' upper floors, including the people who were working there, are scattered on the street. A substantial amount of debris clutters the entire area. Five percent of the population between the five and two psi rings are dead. Forty-five percent are injured.[19]

The radius of the two psi shockwave created by a nine-megaton bomb could extend as far as 17 miles (27 km) depending on local topography and the altitude at which the bomb exploded.[20]

An overpressure of two psi (0.14 kg/cm^2) would produce a force of 288 pounds on every square foot (4600 kg/m^2) of surface exposed to the blast wave. A person 5 ½ feet (167 cm) tall weighing 130 pounds (59 kg) would present a surface area of about 8 square feet (0.74 m^2) if they faced directly toward, or away from, a blast.[21] Exposed to a two psi shockwave, the total force on their body would be about 2300 pounds (3400 kg).

If the blast doors at a Titan II complex were subjected to two psi of overpressure the force on the door would be 12,000 pounds (5400 kg). At their full design-limit overpressure of 1000 psi (70 kg/cm^2) the total force applied to the door would be about 6,000,000 pounds (2,700,000 kg).

Interestingly, the destructive power of a nuclear weapon does not increase in direct proportion to its yield. For example, a 20-megaton bomb would not do twice as much damage as a 10-megaton bomb. Doubling the yield of a given bomb would increase the radius of destruction only by about 25 percent. To double the radius of destruction the yield of the bomb would need to be increased by a factor of eight, from 10 to 80 megatons, for example.[22]

The reasons for this are complex, but in simple terms energy from a nuclear detonation is released in three-dimensions—the explosion is a spherical event. Energy from the bottom half of the sphere is directed mostly into the ground while energy from the top half of the sphere is directed mostly into the sky. It is only the energy in a small band around the equator of the sphere that is useful, if "useful" can be the right word. As the yield of the bomb increases the sphere gets larger, but the band of useful energy increases only slightly.

Another consequence of a nuclear detonation is residual radiation. As mentioned earlier, atoms of heavy elements like uranium or plutonium are split into lighter elements by means of fission, releasing energy in the process. The resulting lighter atoms are referred to as fission products. Since there are about 40 ways to split an atom of uranium or plutonium, a fission bomb will create a total of about 80 different fission products—almost every known naturally occurring element. These new atoms are often created in ways that make them unstable. In an effort to return to a stable condition they "spit out" excess energy in the form of a variety of particles and sometimes gamma rays. These emissions are called radioactivity. In the process of shedding particles, some of the atoms are transformed into yet other elements. These are called decay products, many of which also are radioactive.[23] Radioactivity persists until all the atoms have reached a stable condition—a process that, in some cases, may take tens of thousands of years. It should be noted that the radioactivity produced in the detonation of a hydrogen bomb comes almost entirely from the uranium or plutonium in the small fission bomb used as the detonator, and from uranium used as a casing for the bomb. While fusion generates large quantities of neutrons (not themselves radioactive but which can induce radioactivity in a wide variety of materials) the fusion process itself is quite clean.

Most of the vaporized radioactive fission and decay products are sucked up into the mushroom cloud. As they cool they begin to condense on particles of dust and debris in the cloud. These dust particles are pulled high into the sky by the rising column of hot air and are carried away by the wind. The heavier particles fall out of the cloud soon after the explosion (hence the term "fallout"), while the lighter ones may remain aloft for months depending on their weight. The fallout, much of which is furiously radioactive, will distribute itself downwind of the explosion, gradually settling on the ground tens or even hundreds of miles (km) away.[24] Fallout can kill people far removed from the direct effects of the explosion that caused it, and may persist at dangerous levels for several weeks.[25] Even after radioactivity from fallout has subsided to less than lethal levels, elevated levels of radioactivity would persist for decades.

Radioactivity is dangerous because of its ability to destroy cells in living tissues and cause mutations in cell DNA. If cells are destroyed faster than the body can replace them, sickness or death may follow. Mutations in cell DNA are more insidious in that the damage is permanent and may result in a variety of cancers (which may not manifest themselves for decades), an increase in birth defects, and generally shortened life expectancy.[26]

1 Tsipis, 29-35

2 Paul Craig and John Jungerman, *Nuclear Arms Race: Technology and Society*, (McGraw-Hill, 1986), 144-148

3 http://science.msfc.nasa.gov/ssl/pad/solar/interior.htm

4 Glasstone, 22

5 Isotopes are elements that have too many or too few neutrons. See http://en.wikipedia.org/wiki/Isotope

6 Rhodes, 483

7 Ibid., 482-483

8 The actual length of this hypothetical train varies depending on the assumed size and length of the freight cars.

9 The radius of the 2 psi shockwave would extend about 17 miles from the point of detonation, thus yielding a circular area of destruction of about 900 square miles.

10 Glasstone, 8

11 http://www.fas.org/nuke/intro/nuke/effects.htm

12 Glasstone, 29

13 Ibid., 328-334

14 Ibid., 644

15 Ibid., 29

16 Ibid., 644

17 Ibid., 42-44

18 Ibid., 87-90, 111-115

19 PBS/WGBH *The American Experience*, "Race for the Super Bomb," http://www.pbs.org/wgbh/amex/bomb/sfeature/1mtblast.html

20 Glasstone, 637

21 http://www.medcalc.com/body.html

22 Glasstone, 127-133

23 Ibid., 19-21

24 Ibid., 39-42

25 Ibid., 649-657

26 Ibid., 611-625

▶ Titan II airframes were staged at the Aircraft Maintenance and Regeneration Group (AMARG) in Tucson, Arizona. They went to the shredder in 2009. The long sections are first stages, the shorter ones are second stages. Interstage coupler sections can also be seen.

CHAPTER 10

RUINS

Contrary to popular belief, Titan II's deactivation was not a result of the Strategic Arms Limitation Treaties (SALT). The idea to retire Titan II was, in fact, based largely on economics and a plan to update the nation's nuclear deterrent.

There were many arguments both for and against the deactivation of Titan II. Those in favor of keeping the missiles on alert cited the fact that Titan II's huge warhead was uniquely suited for use against wide-area soft targets

◀ A row of stage-two air frames bakes under the desert sun at AMARG.

◀ After 20 years of service in a darkened silo, a row of RVs show their age after 20 years in the sun. An adapter section can be seen at right.

▲ Stencil on RV ablative skirt.

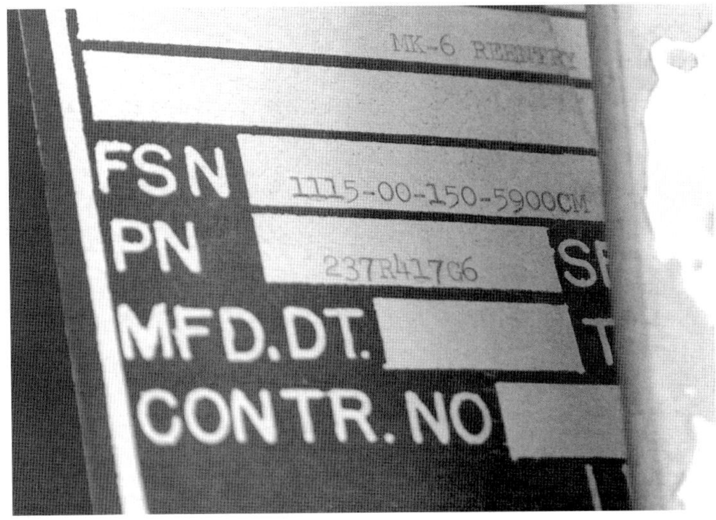

▲ RV serial number plate.

containing multiple "designated ground zeros" (cities with several military and/or industrial facilities such as refineries, for example) and claimed that the missile's absence would increase U.S. vulnerability. Those against pointed to the missile's less-than-stellar safety record together with two serious and high profile accidents. (In fairness it should be understood that Titan II was an extremely reliable machine, and that all accidents involving Titan II could be traced to human error, and not some malfunction of the missile itself). Then, in 1981, the Reagan Administration announced its desire to modernize the country's nuclear deterrent forces. Modernization notwithstanding, in the end it seemed to come down to money. Over the years, inflation and other pressures had pushed the operating costs of Titan II into the millions per year per site, and by the early 80s these costs finally tipped the balance. The Air Force estimated that it could save more than $500 million per year by pulling the plug on Titan II.[1]

Deactivation of the entire program began with the 390th Strategic Missile Wing in Tucson in 1982. The program was named RIVIT CAP. While the code name sounds somehow appropriate for the mission, the name had no special significance and more than likely was picked because it was the next available designation in some big book of Air Force code names.[2] RIVIT CAP proceeded in three phases.

Phase I began with the selected missile being taken off alert status. The first major operation was to remove the warhead and off-load the fuel and oxidizer. Much of the propellant was stored at Aerojet's facilities in California, at Vandenberg AFB and at Cape

•SALE•
TITAN II MISSLE
Dismantlement Project
FUEL TANKS; 60,000, 20,000 & 8,000 Gallon
TOWERS; 30' to 60' Galvanized Steel, Antennas, Windmill, Wind Generator, Ham Operator Antenna
TELEPHONE POLES; Light Poles, Etc.
ELECTRICAL; Transformers, Switch Gear, Outdoor Lighting, Etc.
ELEVATORS; Blowers, Fans & Motors
PUMPS; Sump, Line & Submersible
HYDRAULICS; Pumps, Motors Cylinders, Rams, Valves, Fittings, etc.
STEEL USABLE
PLATE; 3/8, 1, 1½, 3½, & 9½"
I-BEAM — FLANGE BEAM — ANGLE IRON
MANHOLE COVERS & Rings — Grating
PIPE; 2-24" Steel & Aluminum
STAIRS; Rail & Decking
REDWOOD; 1" x 6" Used
MISCELLANEOUS; Meters, Gauges, Temp. Thermometers, Brass & Stainless Steel Valves — Air Receiver Tanks, Speakers & Horns — Roof Jacks — Expanded Metal Panels, Etc.
FOR INFORMATION CALL
LITWAK-GRACON SALES
TUCSON, AZ 602-887-1487

TMM ARCHIVES

▶ Everything that could be salvaged, was. Advertisement in Arizona Daily Star (Tucson) newspaper. Date unknown.

USAF PHOTO, TMM ARCHIVES.

▼ To salvage the generator, contractors first cut away the launch duct's steel plating, then used high explosives to breach the wall. The generator was protected from the blast with a temporary wall constructed of heavy wooden beams.

▶ The 14-ton generator dangles from a crane after being lifted out of the silo. Salvage of the generator may have been done only at Tucson as the process proved both expensive and time consuming.

USAF PHOTO, TMM ARCHIVES.

▶ More than seven hundred tons of high quality steel made it worth cutting up the silo door and hauling away the pieces. Here, pieces of the door are being loaded onto trucks.

USAF PHOTO, TMM ARCHIVES.

USAF PHOTO, TMM ARCHIVES.

◀ Not very much was left behind. The 26,000-pound thrust mount was pulled out, cut up, and carted away.

USAF PHOTO, TMM ARCHIVES.

◀ This photo looks down the silo to level one. The box-girder has been cut away revealing the very top-most sections of the launch duct wall (center and lower left), as well as the silo door motor and the top of the silo elevator shaft. Also seen are the interior of the north exhaust shaft (right) and the acoustic modules lining the shaft wall.

▲ After 20 years of service, it was all over in less than a second. About 2800 pounds of high explosives made short work of the top 25 feet of the silo.

BOTH: TMM ARCHIVES.

USAF PHOTO. TMM ARCHIVES.

▲ The reinforced concrete silo cap.

Canaveral for future use in Titan II as a space launch vehicle. When finding adequate storage facilities became a problem some of the oxidizer was sold to industrial users, while some of the fuel was simply burned.[3]

Once the propellant had been off-loaded, the missile itself was removed and shipped by rail for storage at Norton AFB, near San Bernardino in southern California. When Norton was closed, the missiles were transferred to the Aerospace Maintenance and Regeneration Group (AMARG), Tucson's famous aircraft "boneyard" at Davis-Monthan AFB, where they were shredded in 2009. The reentry vehicles were stored at two locations: Nellis AFB near Las Vegas, Nevada, and Kelly AFB near San Antonio, Texas. Eventually, several RVs made their way to AMARG, were they still reside as of this writing. The Titan Missile Museum holds three in its collection, two of which are on display. The actual bombs are presumed to have been returned to Pantex, America's only nuclear weapons assembly and disassembly facility, near Amarillo, Texas.

With the missile gone, all of the documents in the red emergency war orders safe were removed and returned to the wing command post. At that point, the crew no longer needed to stay at the complex 24 hours a day.

Once all the various missile and ground support systems had been shut down, private contractors took over the process of dismantlement (Phase II). As the phase-out of all 54 sites would take several years to complete, spare parts were still required. Contractors therefore removed any equipment the Air Force wanted to keep for use at sites that were still active, as well as for potential use elsewhere. The list included the large backup generator. Removal of the generator required blasting a hole in the 18-inch (46 cm) steel-plated concrete launch duct wall and pulling the unit out with a crane.[4] When the Air Force had everything it wanted, the contractors got salvage rights to everything else. This included literally miles of copper wire, cables and piping, pumps, motors, lights, and whatever else was thought to be worth selling on the surplus or scrap metal market.

Once salvage operations had been com-

pleted, demolition of the silo (Phase III) began. Of course, destroying a complex that was built to withstand a nuclear explosion is exceedingly difficult. But complete destruction of the complex was unnecessary. The only part that really counted, at least as far as the Soviets were concerned, was the launch duct, destruction of which made the rest of the complex irrelevant. That was a good thing in the eyes of the Air Force, which wanted to spend as little as possible to get the job done to the satisfaction of applicable treaties.[5] It should be noted that while Titan II was not deactivated because of a treaty, existing treaties did require dismantled sites to be destroyed in a manner that would render them forever unusable.

The mammoth silo closure doors were cut into small pieces and carted away as scrap metal. As scrap, each door was worth about $37,000 in the dollars of the day, or around $70,000 in 2007 dollars.[6] The thrust mounts were taken as well. As for the task of destroying the silo itself, an engineering expert once quipped, "There is no problem that can't be solved with the suitable application of explosives." Demolition crews used more than 2800 pounds (1300 kg) of high explosives to blow up the top 25 feet (8 m) of the silo. Old tires and steel netting laid on top of the silo apron were used to help contain the explosion, and it was a sight to see.

After the "big bang" the sites were left untouched for 30 days while Soviet satellites verified that the complexes had been destroyed. Then the debris generated by the explosion was plowed into the launch duct. The remaining depth of the launch duct was filled with whatever detritus was close at hand, and the ruins were capped with a reinforced concrete disk 60 feet (18 m) in diameter and 3-feet (1 m) thick.

Army Corps of Engineers' drawings detailing the demolition process indicate that the long cableway tunnel joining the silo to the blast lock was to be abandoned.[7] But in at least two cases, excavation by the new owners revealed that the long cableway had been dug up and removed.[8] In one other case, investigation with a magnetometer suggested the cableway may still be present.[9] No specific information regarding cableway status exists for other sites near Tucson, but archival photographs support the idea that most or all were removed. And why not? One hundred and forty feet (43 m) of half-inch thick (1.3 cm) steel pipe 9 1/2 feet (2.9 m) in diameter—and the massive shock-isolated frame within—was worth a small fortune as scrap metal. Nevertheless, digging up the cableways may have cost almost as much as the pipes were worth. While no conclusive proof

◄ 570-3 is located within the Ironwood Forest National Monument, west of Tucson, Arizona, and can be legally visited. 0-3 is typical of launch complex ruins. The buried silo cap lies between the fuel and oxidizer hardstands. Antenna artifacts can be seen in the vicinity of the control center.

▲ If something was too difficult to remove it was abandoned in place. The fuel and oxidizer hardstands were two such objects, and are common to virtually all launch site ruins.

exists as of this writing, there are suggestions that the cableways may have been salvaged only at Tucson.

The access portal elevator and stairway were removed, and additional high explosives were used to blast off the roof of the structure. The empty shaft was then filled with dirt and rubble. The airshafts of the silo and LCC were filled with grout, as were the silos containing the HF and IRCS hardened antennas, and manhole J. Artifacts on the surface that were easy to remove (like the various poles for lights and sirens) were taken out, and the surface of the complex was graded to restore natural contours. The sites were then sold at auction, mostly to private individuals who bought them primarily for the ten acres of land. It is likely that most buyers had little or no understanding of what lay beneath their newly acquired property.

Indeed, beneath all 54 sites, the blast lock and LCC, sans equipment, lay fully intact. And while the launch duct had been filled in, much of the silo equipment area surrounding it remained, though perhaps somewhat the worse for wear from the explosion.[10]

As the years ticked by, commercial development that had once encroached upon the launch sites began to cover them over, while the memory of what lay underground continued to fade. A restaurant was built on top of one site, a garden center on another, and in perhaps the most ironic turn of events, a church was built on a third. Still, most lay vacant and were slowly reclaimed by the desert.

It is important to understand that all Titan II

▲ The top of the HF hardened antenna silo is easy to recognize. The small square at left is the base of what was once the mooring for a block and tackle used to remove the blast plug from the access hatch.

▲ The IRCS hardened antenna silos are identifiable by the outline of the hatch. The two antennas were generally within a few yards (meters) of each other.

▲ Like the antenna silos, the airshaft was filled with grout. Note protruding left and right sides of the escape ladder.

sites, with the exceptions of 571-7 and 570-3, are now owned by private individuals or local governments and thus are private property. Many are posted "no trespassing" and those wishing to visit a site are strongly advised to visit only the Museum and 570-3.

Most of the artifacts seen at abandoned sites today are those concrete structures that were too difficult to remove. Most obvious among these are the fuel and oxidizer hardstands. These large rectangular platforms served as parking spaces for propellant tank trucks. The concrete cap placed over the silo is buried deep below the surface, but its location can be determined with some degree of accuracy by imagining lines running perpendicular to the center of the hardstands. When extended, these lines cross very near the center of the silo.

Other artifacts include the tops of the silos for the HF hardened and IRCS pop-up antennas. These disks of concrete, 17 feet (5 m) in diameter, are generally found 300 feet (90 m) or so west southwest of the silo. The IRCS silo tops are easy to recognize as the outline of the hatch has a distinctive shape.

The base for the UHF antenna, a concrete disk 10 feet (3 m) in diameter and four feet (1.2 m) thick, has not been identified among the ruins of several sites visited by the author.[11] Investigations have confirmed that at two sites, contractors disposed of the UHF antenna base by dropping it into the access portal.[12] This is likely to have been done at most other sites as well.

▲ Most sites got their water from wells, and some sites had more than one.

The base for the IRCS soft antenna is visible at most sites. It is a square concrete pad with bolts embedded near each corner. These bolts secured a metal tower to the base.

The top of the LCC lies hidden about nine

▶ 571-4 provides an absolutely unique perspective of a Titan II launch complex. Sufficient dirt has been excavated to expose portions of the once-buried structures. At right is the access portal. The top was originally level with the surface of the ground. The dome at left is the top of the control center, once nine feet below the surface. The tops of the two IRCS hardened antenna silos and manhole J are seen behind. Caution: No trespassing.

◀ Typical abandoned LCC. HF/UHF radio cabinets at left. The SLFCS cabinet is gone. ALOC and FPCB cabinets are at right. The LCCFC would have been in the center. It is pitch dark. The only illumination in this image was the camera's flash.

feet (2.7 m) below the surface at most sites, but its location can be estimated if the airshaft that supplied fresh air for the crew can be found. The airshaft was a three-foot (1 m) diameter pipe that terminated at ground level in the center of an eight-foot (2.4 m) square concrete pad. Looking northeast across this square one looks across the LCC, past the blast lock, and down the long cableway toward the silo. The center of the LCC would be about 30 feet (9 m) from the airshaft. The area of the blast lock between blast doors 8 and 9 lies at a point some 50 feet (15 m) further. About 30 feet (9 m) southeast of that point lies the center of the access portal, although the ruins of the portal itself are not generally visible.

Wandering around a complex one might also find the remnants of the well, the edges of the soft water storage reservoir, pieces of rebar and wire poking out of the ground, and literally dozens of other bits and pieces of what was once the most powerful missile weapon system ever deployed by the U.S.

One of the most remarkable launch complex ruins is that of site 571-4, southeast of Tucson, Arizona. The complex was purchased by Pima County and its soil was used as a source of high quality fill for road projects, among other things. As of this writing the site is apparently in disuse, but so much dirt was excavated that portions of the once buried complex are now visible. The top of the LCC dome for example, once nine feet (2.7 m) underground, now is ten feet above the ground. More than ten feet (3 m) of the IRCS pop-up antenna silos are exposed. Manhole J now stands as a huge concrete cube on the surface. The top of the access portal, originally at ground level now towers over your head. These artifacts enable easy location of the still submerged blast lock and silo.

Recently, more than 20 years after the last complex was destroyed, a few ambitious owners have set out on the adventure of excavating the sites under their property. At least one verified account exists of someone jack-hammering his way down through the grout-filled LCC airshaft (a task complicated by having to get past either the emergency escape hatch or the LCC blast valve). However, most others who have gone through the considerable effort and expense

◂ LCC level 3. Empty communications equipment racks at left with battery power supply cabinets at right.

of re-entering their buried LCCs have done so by digging out the access portal and opening blast door 6. In every case, this has proven to be a laborious, expensive, time-consuming and dangerous proposition.

What makes excavation of the portal so incredibly daunting is that demolition contractors used it as a convenient means of burying all sorts of debris. Those who have excavated the portal have encountered numerous chunks of concrete weighing thousands of pounds, twisted rebar, large metal fragments and even the smashed remains of the portal stairway. In two cases the base of the UHF antenna, a concrete disk weighing some 44,000 pounds (20,000 kg) has been found in the portal, requiring a special crane to remove.

Once past the rubble of the access portal, one is confronted with the blast doors. In some cases these have been found to be closed but unlocked. In other cases they have been found wide open. But there is always the chance, as in one documented case, that one or more of the doors will be found welded shut. In that case, extra time, money and effort, and perhaps special equipment, would be required to gain entry to the LCC.

Launch control centers that have been accessed have been found to be in generally good condition. All equipment has been removed, though reports exist claiming that in a few cases some equipment was left behind. Generally, most or all of the empty equipment racks and cabinets are found in place.

While sites around Tucson have been found to be dry, at Little Rock and Wichita the possibility that the underground structures have been flooded cannot be ignored. Flooded sites could be pumped dry, cleaned up, and maintained with sump pumps—all at additional cost, of course.

Clearly, the excavation of a Titan II launch control center is not a task to be taken lightly. Very significant expenditures of time, effort, and money are required to do the job right, and safely.

1 Stumpf, 253-254

2 LtCol. Orv Doughty, USAF (Ret), TMM docent training session, 1999

3 Stumpf, 255-256

4 Ibid. 260

5 Ibid, 260

6 United States Department of the Interior, U.S. Geological Survey report, *Metal Prices in the United States through 1998* (http://minerals.usgs.gov/minerals/pubs/metal_prices/)

7 Stumpf, 259

8 Author's research.

9 Property owner's investigation

10 Author's research

11 Ibid.

12 Ibid.

CHAPTER II

EPILOGUE

From their beginnings as simple pyrotechnics to their ascent as complex machines that carried people into space, rockets and missiles have proven their usefulness in a wide range of applications. We have used them to explore the Earth, the Moon and the farthest reaches of the solar system. Rockets have lofted satellites into orbit providing worldwide communications and weather forecasts. But the same technology that conveyed men to the moon has also made possible the conveyance of destructive power on a scale that defies both description and imagination, and made it deliverable anywhere on Earth in about the same time it takes to deliver a pizza to your doorstep.

With the invention of atomic weapons in 1945, both the United States and the former Soviet Union scrambled to build their nuclear arsenals. The resulting "cold war" and its ensuing "arms race" evolved into a feverish 40 year effort to maintain a "balance of terror:" an equilibrium in which each side was capable of annihilating the other—but at the expense of being annihilated in the process. Titan II was born of America's deepest fears: not only that a nuclear exchange might end our way of life, but also that it might end the entire world as we know it. The fear of mutual annihilation, combined with the knowledge that there is no defense against nuclear weapons, became a powerful deterrent force. A nuclear war would be a war no one could actually win.

The good news is that the concept of deterrence, as delivered by weapon systems like Titan II and its successors, have kept the nightmare of nuclear war at bay for more than 50 years. The bad news is that this same brand of deterrence may have little influence on a new class of enemies here in the 21st century, many of whom may be seeking nuclear weapons of their own.

If nuclear war is a messy business, its prevention may be no less so. In his 1979 book, *Nuclear Nightmares*, Nigel Calder wrote, "We all live in a house of cards, and any rearrangement of the pieces, however sensible in the long run, could bring everything crashing down." He was referring to the seemingly insurmountable difficulties encountered in attempting to back away from nuclear weapons through reductions or disarmament, whether negotiated or self-imposed. Added to those difficulties today are the lack of a clearly defined enemy, the proliferation of nuclear technology to rogue states or groups, and the fact that while Titan II is gone, the world still harbors thousands of nuclear weapons, aimed and ready to launch at a moment's notice.

In the future, preventing a nuclear conflict will require more than deterrence. It will require a clear understanding of the history of conflict, the reasons governments choose to take up arms and the consequences of those choices.

Some form of deterrent may always be essential, but in the end the military does only what its civilian superiors tell it to do. That would be you.

Titan II lived through some of the darkest moments of the Cold War, and has a great deal to tell you. Listen carefully. As the poet and philosopher George Santayana once remarked, "Those who do not understand the past are condemned to repeat it."

LGM-25C (TITAN II) SPECIFICATIONS

AIR FRAME

Stage One

Airframe Manufacturer	Martin Company
Length (including interstage section and engines)	70.17 ft (21.38 m)
Weight (with engine, less propellant)	9583 lbs (4346 kg)
Weight (with engine and propellant)	Nominally 255,035 lbs (115,682 kg)
Diameter (excluding conduits and air scoops)	10 ft (3.04 m)
Engine Model	LR87-AJ-5
Engine Manufacturer	Aerojet General
Engine Weight (dry)	3771 lbs(1710 kg)
Propellant type	Liquid
Fuel Tank Readiness Pressure	9.5-15.5 psig (0.66-1.08 kg/cm^2)
Oxidizer Tank Readiness Pressure	9.5-13.5 psig (0.66-0.94 kg/cm^2)
No. Thrust Chambers	2
Expansion Ratio:	8:1
Thrust Chamber Pressure	795 psia (56 kg/cm^2)
Thrust (Sea Level)	430,000 lbs (195,000 kg)
Burn Time (nominal)	150 seconds
Fuel	Aerozine 50 (N_2H_4 and $(CH_3)_2N_2H_2$)
Fuel Quantity (nominal)	11,600 gal (43,900 l)
Oxidizer	Nitrogen Tetroxide (N_2O_4)
Oxidizer Quantity (nominal)	13,900 gal (52,600 l)
Thrust Chamber Mixture Ratio (oxidizer:fuel)	2.01:1
Total Flow Rate (oxidizer plus fuel)	1639 lbs/sec (743 kg/sec)

Stage Two

Airframe Manufacturer	Martin Company
Length (less RV adapter)	19.26 ft (5.87 m)
Diameter	10 ft (3.04 m)
Weight (less propellant and RV)	5099 lbs (2313 kg)
Weight (with propellant, less RV)	Nominally 63,689 lbs (28,889 kg)
Engine Model:	LR91-AJ-5
Engine Manufacturer	Aerojet General
Engine Weight (dry, including ablative skirt)	1367 lbs (620 kg)
Propellant type	Liquid
Fuel	Aerozine 50 (N_2H_4 and $(CH_3)_2N_2H_2$)
Fuel Quantity (nominal)	2860 gal (10,800 l)
Oxidizer	Nitrogen Tetroxide (N_2O_4)
Oxidizer Quantity (nominal)	3170 gal (12,000 l)
Fuel Tank Readiness Pressure	33.2-40.9 psig (2.32-2.86 kg/cm^2)
Oxidizer Tank Readiness Pressure	36.2-43.9 psig (2.53-3.07 kg/cm^2)
No. Thrust Chambers	1
Expansion Ratio:	49.2:1
Thrust Chamber Pressure	827 psia (57.89 kg/cm^2)

Thrust (Vacuum) .... 100,000 lbs (45,360 kg)
Burn Time (nominal) .... 180 seconds
Thrust Chamber Mixture Ratio (oxidizer:fuel) 1.88:1
Total Flow Rate (oxidizer plus fuel) .... 318 lbs/sec (144.24 kg/sec)
Vernier Manufacturer .... Thiokol
Vernier Motors 2
Vernier Type .... Solid fuel
Vernier Thrust .... 950 lbs total (431 kg total)
Vernier Burn Time .... Up to 20 seconds
Total Length of Missile (with engine and RV) ..... 103.4 ft (31.52 m)
Total Weight of Missile (with propellant and RV) .. Nominally 327,104 lbs (148,372 kg)

GUIDANCE

Manufacturer .... Delco Electronics
Type .... All inertial
CEP .... Less than 1 nautical mile (1.15 miles) (1.85 km)

REENTRY VEHICLE

Manufacturer .... General Electric
Model .... Mark VI
Type .... Ablative
Heat Shield Material .... GE Series 100 plastic
Nose Cap Material .... GE 66-Nylon cloth and phenolic resin
Diameter (at base) .... 7.5 ft (228.60 cm)
Length (less adapter section) .... 10.17 ft (3.10 m)
Adapter Length .... 3.77 ft (114.90 cm)
Weight (less warhead and adapter/decoy section) .. 1375 lbs (624 kg)

WARHEAD

Model W-53
No. of Warheads 1
Type .... Thermonuclear
Length .... 8.66 ft (264 cm)
Diameter .... 36.5 in (92.71 cm)
Weight .... 6200 lbs (2800 kg)
Yield .... 9 megatons
Detonation mode .... Air or ground burst (selectable)

MAXIMUM RANGE FLIGHT PROFILE

	Time	Altitude		Range		Velocity	
	(min:sec)	Mi	(km)	Mi	(km)	MPH	(km/h)
Liftoff	0	0	(0)	0	(0)	0	(0)
End of vertical flight	00:15	0.25	(0.40)	0	(0)	117	(188)
Staging	02:27	47	(76)	47	(76)	5630	(9061)
Stage II cutoff	05:27	217	(349)	442	(711)	15,108	(24,314)
Vernier cutoff	05:43	238	(383)	499	(803)	15,112	(24,320)
RV separation	05:46	242	(389)	513	(826)	15,111	(24,319)
Apogee	19:25	797	(1283)	3108	(5002)	12,598	(20,275)
Reentry	35:14	56	(90)	6192	(9965)	15,968	(25,698)
Impact	36:31	0	(0)	6332	(10,190)	457	(735)

Sources:

Aerojet Liquid Fuel Rocket Company, *Titan II Propulsion Subsystem*, 1981

The Martin Company, *Detailed Design Specifications for Model SM-68B*, 1-September-1961, updated through Change Notice 66, 24-January-1969

TRW Space Technology Laboratory, *Titan II Weapon System Final Report*, January 1965

U.S. Air Force, *Technical Order 2K-LR87-16* (Field Maintenance for Rocket Engine LR87-AJ-5), 1983

U.S. Air Force, *Technical Order 2K-LR91-16* (Field Maintenance for Rocket Engine LR91-AJ-5), 1983

David Stumpf, *Titan II: A History of the Cold War Missile Program*, 2000

GLOSSARY

AAS...... azimuth alignment set
ABM ..... anti-ballistic missile
AC....... air conditioner
AC....... alternating current
ACP...... alternate command post
AFB...... air force base
ALOC .... alternate launch officer's console
AM ...... amplitude modulation
AMARG .. Aerospace Maintenance and Regeneration Group
APS...... accessory power supply
AW ...... air washer
BFRC .... big freaking red cloud
BLX...... basic launch complex
BMAT .... ballistic missile analyst technician
BV....... blast valve
BVL...... butterfly valve lock
BVLC .... butterfly valve lock control
C .... centigrade
CB....... circuit breaker
CB ...... citizen's band (radio)
CBR...... chemical, biological and radiation (filter)
CMG ..... control monitor group
DC....... direct current
DMCCC .. deputy missile combat crew commander
CEP...... circular error probable
EAM ..... emergency action message
ELU...... emergency lighting unit
EMP ..... electromagnetic pulse
ERCS .... emergency rocket communications system
EWO ..... emergency war orders
F .... Fahrenheit
FAC...... failure of alternating current
FM....... frequency modulation
FPCB .... facilities power control board
GMT ..... Greenwich Mean Time
Ha .... Hectares (land area)
HF....... high frequency
HS....... hydraulic system
ICBM..... intercontinental ballistic missile
IMU...... inertial measurement unit
IRBM..... intermediate range ballistic missile
IRCS ..... intercomplex radio communications system
ITU ...... International Telecommunications Union
Kg .... kilogram
kHz ...... kilohertz
kt........ kiloton (one thousand tons)
LCC...... launch control center
LCCFC ... launch control complex facilities console
LOX...... liquid oxygen
MCC ..... motor control center
MCCC.... missile combat crew commander
MFL ..... missile fault locator
MFT ..... missile facilities technician
MGACG .. missile guidance and alignment checkout group
MGC ..... missile guidance computer
MHz ..... megahertz
MSA ..... Mine Safety Appliance
MT....... megaton (one million tons)
NASA .... National Aeronautics and Space Administration
OGE...... operating ground equipment
PDC...... power distribution and control
PSI ...... pounds per square inch
PTPMU... propellant tank pressure monitoring unit
PTS...... propellant transfer system
PTT...... push to talk
RAM ..... random access memory
RFHCO ... rocket fuel handler's clothing outfit
RP-I...... a highly refined kerosene used as rocket fuel
RPIE ..... real property installed equipment
RPM ..... revolutions per minute
RTMN.... radio type maintenance net
RV....... reentry vehicle
SAC...... Strategic Air Command
SAC-CEM. Strategic Air Command civil engineering manual
SACCS ... strategic automated command and control system
SALT..... Strategic Arms Limitations Treaty
SLFCS ... survivable low frequency communication system
SMS ..... strategic missile squadron
SMW..... strategic missile wing
SSB...... single sideband
T.O. ...... technical orders
U.S.S.R. .. Union of Soviet Socialist Republics (the Soviet Union)
UHF...... ultra high frequency
UTC...... coordinated universal time
VDAP .... vapor detector annunciator panel
VIP ...... very important person
VLF...... very low frequency
VSS...... voice signaling system
WCP ..... wing command post
WCT ..... wire control and transmission
WTMN ... wire type maintenance net
WWV .... radio callsign for the National Institutes of Standards and Technology shortwave time station
ZULU .... military jargon for UTC

BIBLIOGRAPHY

Some of the books listed in this bibliography have been out of print for a number of years. Most of these were written in the 70s and 80s—while Titan II was still operational—and provide an exceedingly helpful context useful in a more complete understanding of the Cold War and nuclear weapons in general, and Titan II in particular. With very few exceptions these works are neither rare nor expensive, and can easily be obtained from Amazon.com or other Internet book sellers.

Clarfield, Gerard, and William Wiecek, **Nuclear America: Military and Civilian Nuclear Power in the United States, 1940-1980.**
Harper and Row, 1984, 518 pages.
ISBN: 0-06-091244-8

A thorough, well researched, and engaging study of the politics surrounding the development and use of nuclear power and nuclear policy. In addition to a discussion of the obvious military applications of nuclear power, Clarfield and Wiecek also provide insight into the development of civilian nuclear power and its complex interrelationship with the military. The book recounts the entire evolution of policy from Eisenhower's "massive retaliation" and Kennedy's "mutual assured destruction," to T. K. Jones (a mid-level Reagan administration official) who assured us that we could survive a nuclear attack by digging holes in our backyards and driving our cars over them. From "no nukes" to Three Mile Island this is a fascinating look at the complex and ever-changing landscape of nuclear policy. It also contains a thoughtful and exceptionally even-handed discussion of the decision to use the Bomb on Japan near the end of World War II.

The Effects of Nuclear War.
U.S. Office of Technology Assessment, 1979.
Public domain. Gale Research Company published a copyrighted compilation in 1984.

Originally written in 1979 at the behest of the U.S. Senate Committee on Foreign Relations, the book was updated in 1984 to include a number of working papers intended for a never-published second volume. The book attempts to describe the effects of hypothetical nuclear attacks on Detroit and Leningrad, and presents several scenarios ranging from a limited attack to an all-out exchange involving a variety of civilian, military, and industrial targets. Each case examines the short and long-term effects on civilian populations, economies, and society at large. Included in the book is a work of fiction commissioned by OTA intended to provide a more concrete understanding of the situation in which survivors of a nuclear war would find themselves. This story describes the aftermath of an attack on Charlottesville,

Virginia. The story does not suppose martial law, or that society has descended into anarchy, and paints a picture that is grim but probably overly optimistic. Nevertheless, it makes for an interesting and thought-provoking read.

Emme, Eugene, ed.,
The History of Rocket Technology.
Wayne State University Press, 1964, 320 pages.

A collection of essays by a variety of people involved in various aspects of rocket development. This book contains works by the likes of G. Edward Pendray, who worked with Robert Goddard, and Walter Dornberger, the chief of the German Army Board of Ordnance during WWII and the man who hired Wernher Von Braun, who also wrote a chapter for the book. While some chapters, like Robert Perry's treatise on Atlas, Thor, and Titan are exceedingly well done, others, like Pendray's, suffer from a lack of objectivity because of his close personal relationship with Goddard. Pendray paints an uncritical portrait of Goddard, and Dornberger's piece on the V2 makes no mention of the horrific slave labor used to build the missile. Still, there is a lot of very useful and detailed information, and as the book was written in 1964—while the smoke was still clearing from the launching pads—it would be hard to find authors who were closer to the subject than these.

Ford, Daniel,
The Button: The Pentagon's Command and Control System-Does it Work?
Simon and Schuster, 1985, 270 pages.
ISBN: 0-671-50068-6

A sobering examination of the nuts and bolts of the communications systems designed to send commands to war-fighting commanders on land, in the air and at sea, as these systems were configured prior to 1985 (during the tenure of Titan II). Ford contends that the vast majority of these systems would have collapsed very rapidly in the opening rounds of a nuclear exchange leaving troops in the field in the dark and unable to use their missiles or other weapon system. In addition, Ford claims the emergency communications lines between the U.S. and the Soviet Union would also have broken down, precluding any negotiations that could stop the exchange, or at least prevent the war from escalating. Ford paints a disquieting portrait of a frail command and control system unable to withstand EMP, blast effect or acts of terrorism, and held together with what amounted to baling wire and twine. We can only hope things have improved.

Glasstone, Sam, ed.,
The Effects of Nuclear Weapons.
Department of the Army Pamphlet Number 39-3, Headquarters: Department of the Army, 1962, 730 pages. Public domain.

This book is a U.S. Army publication and it has been revised several times. The first edition was published in the late 50s with updates every few years until at least the late 70s. It is a technical, but readable, volume including information on air blast phenomena; air blast loading and target response; expected structural damage effects; air, surface and subsurface effects; radiation emissions and effects; fallout; radio and radar effects; effects on personnel and strategies for protection, among other things. The 1962 version contains 730 pages loaded with charts, diagrams, and formulas. The book also included a circular plastic slide-rule "nuclear bomb effects computer" (often long since lost) that permits rapid calculation of overpressure, yield, range of effects, crater and fireball size, wind velocities and other phenomena associated with nuclear detonations. Glasstone's book is regarded as the last word on the effects of nuclear weapons and is a must-have for any serious student of the subject.

Hansen, Chuck,
U.S. Nuclear Weapons: The Secret History.
Orion Books, 1988, 232 pages.
ISBN: 0-517-56740-7

In spite of its somewhat sensationalist title, Hansen's book is the single most authoritative unclassified document available on nuclear weapons developed by the United States. It is a highly detailed description of virtually every warhead ever assembled (up to 1988), gathered entirely from unclassified (or declassified) sources. From the MK 1 to the MK 105 and from the W 1 to the W 91, Hansen includes information on yield, inventory, dates of deployment and retirement, and how the devices were deployed. The book is full of high quality photos and diagrams. Hansen starts with a superb though somewhat technical discussion of the physics of nuclear weapons, then moves on to a finely crafted summary of nuclear weapons development, and an entire chapter on the development of thermonuclear weapons and delivery systems including gravity bombs, artillery shells, and missiles. Long out of print, your best bet for finding this marvelous book will be on the Internet, but be cautioned—it will be expensive, typically running $200-300 per copy. In 1995 Hansen re-wrote the book and offered it as an eight-volume set of CDs titled *The Swords of Armageddon.* The set is still available ($300) from www.uscoldwar.com.

Isaacs, Jeremy and Taylor Downing,
Cold War: An Illustrated History, 1945-1991.
Little, Brown and Company, 1998, 438 pages.
ISBN: 0-316-43953-3.

Released as a companion to the CNN TV series. The medium of television, when artfully done, has a way of distilling and explaining a subject in a way that no other medium can. So it is not surprising that the producers of CNN's series on the Cold War have done such a nice job in transferring their work into print. Isaacs and Downing have organized the book into 20 roughly chronological chapters from the rise of the "iron curtain" to the fall of the Berlin Wall, covering all the events and individuals figuring prominently in the conflict. The book is unusually approachable because of its genesis in television and is profusely illustrated with hundreds of (mostly black and white) historic photos.

Launius, Roger, and Dennis Jenkins, ed.,
To Reach the High Frontier: A History of U.S. Launch Vehicles.
University of Kentucky Press, 2002, 519 pages.
ISBN: 081312245-7

The title says it all. From Robert Goddard to Wernher von Braun, and from the V-2 to the Saturn V, with stops for Sputnik and the Space Shuttle along the way. *To Reach* is a reasonably thorough discussion of the development of civilian and military launchers written in a not-too-technical style. Each chapter was written by a different author, so the style and clarity varies somewhat. This author noted a few factual errors and a couple of mislabeled charts in the chapter dealing with Titan II, but overall, the work seems accurate and authoritative.

Lonnquest, John, and David Winkler,
To Defend and Deter: The Legacy of the United States Cold War Missile Program.
Defense Publishing Service, 1996, 605 pages.
Public domain.

This encyclopedic volume was never widely available and was probably distributed only to selected government and military officials. The book is somewhat less inclusive than its title suggests as the authors focus only on land-based missiles. Nevertheless, their discussion takes in IRBMs, ICBMs, and ABMs, and their associated facilities. These include Thor; Jupiter; Atlas; Titan I; Titan II; Minuteman I, II, and III; Snark; BOMARC, and the entire Nike series, all complete with technical specifications and basing plans. Several other systems are mentioned in passing. The authors also provide one of the best and clearest discussions of the history of U.S. missile development to be found. Lots of great, crisp black and white photos, charts and diagrams. You'll have to search online book sellers to find this book.

Mann, Robert,
Complete Idiot's Guide to the Cold War.
Alpha, 2002, 359 pages. ISBN: 002-864246-5

It's hard to compress the history of the Cold War into 350+ pages still, Mann does a respectable job in hitting the "highlights" and provides enough background to give the reader an acceptably complete overview. From communism to McCarthy, and from the Cuban missile crisis to glasnost (and written just after the September 11 attacks), it's all here, in abbreviated form at least. Like all idiot-type books, this one should not be regarded as complete or entirely accurate, but it is sufficient to help the reader at least appear to understand the subject. Written in short, easy to read segments, it also contains lots of quick facts and recommendations for further reading.

Neufeld, Jacob,
Ballistic Missiles of the United States Air Force: 1945-1960.
Office of Air Force History, 1989, 409 pages.
ISBN: 0-912799-62-5. Public domain.

There may be no better or more authoritative book on the subject that this one. In seven brilliant chapters, Neufeld covers the entire story of ballistic missile development starting from its birth after WWII, to its growing pains of the early 50s, and its culmination as an operational force in 1959. To say the book is thorough is an understatement. The number of missiles included requires over two full pages in the index. From obscure systems like Able, Wizard, Snark and Sparrow, to the more familiar Atlas, Thor, Titan and Minuteman, virtually every missile ever produced (up to 1960) is covered, at least in passing. In addition to the technical details, which are treated in understandable terms, Neufeld devotes considerable time to the politics and inter-service rivalry that both helped and hindered development. If you wanted only one book on the history of ballistic missiles, this would be it.

O'Keefe, Bernard,
Nuclear Hostages.
Houghton Mifflin Company, 1983, 252 pages.
ISBN: 0-395-34072-1

As an ensign in the U.S. Navy, O'Keefe worked on the Manhattan Project helping to design the firing circuits for the atomic bomb. Later, as a civilian contractor with EG&G, he was involved extensively in the testing of nuclear weapons. In *Nuclear Hostages*, O'Keefe begins with a highly distilled discussion of the various discoveries that lead to the Bomb, then, in somewhat greater detail, he recounts his involvement with the Bomb and its many subsequent tests. Perhaps the most interesting chapter is Bravo, O'Keefe's riveting firsthand account of the test of a thermonuclear bomb at Bikini Atoll as experienced from his vantage point at the control bunker 20 miles away. The Bravo shot was predicted to have a yield of about 7 megatons. The actual yield was more like 15.

Preston, Diana,
Before the Fallout: From Madam Curie to Hiroshima.
Walker and Company, 2005, 400 pages.
ISBN: 0-8027-1445-5

This is not a book about the atomic bomb. This is a book about the discoveries that made the atomic bomb possible. Preston recounts Henri Becquerel's observation of mysterious emanations from uranium in 1886 and how it touched off a flurry of research that, in time, unveiled the inner workings of the atom and lead to the discovery of the electron, proton and neutron, and ultimately to the discovery of fission, and the nuclear chain reaction. This is a well-written account of the scientific and political steps, missteps and intrigue surrounding the pivotal events leading to the idea of an atomic bomb.

Rose, Kenneth,
One Nation Underground.
New York University Press, 2001, 313 pages.
ISBN: 0-8147-7522-5

An exploration of the fallout shelter and its place in civil defense, and why it never really caught on. Rose demonstrates that the seemly simple fallout shelter had very complex ramifications for U.S. nuclear and fiscal policy: Would it send the Soviets a message of our resolve, or would they think we were preparing for an attack? And how much was the government willing to pay to provide shelter for its citizens? There were more philosophical questions as well: Would fallout shelters cause politicians to think a nuclear war was winnable? Would a post-nuclear life be worth living? Or would the living envy the dead? And there is some welcome comic relief in Rose's dissertation as he points to some of the more patently absurd facets of civil defense. Duck and cover!

Stumpf, David,
Titan II: A History of the Cold War Missile Program.
The University of Arkansas Press, 2000, 320 pages. ISBN: 155728-601-9

David Stumpf's masterful book is regarded as the definitive history of Titan II, its development and deployment. In ten well-written chapters the author examines the origins of the U.S. ballistic missile program with a brief discussion of Atlas and Titan I, then goes on to fully explore the rise, operation, and eventual fall of Titan II. In addition to technical details, Stumpf also provides a glimpse into the human side of ballistic missiles by devoting two entire chapters to the people who lived, and occasionally died, with Titan II. Stumpf interviewed dozens of people for the book and had access to thousands of documents from military and industrial sources.

Tsipis, Kosta,
Arsenal: Understanding Weapons in the Nuclear Age.
Simon and Schuster, 1983, 342 pages.
ISBN: 0-661-44073-X

This is an exceptionally complete and highly readable book designed for the layperson. It covers everything from the discovery of nuclear energy to the development of nuclear weapons, and includes chapters devoted to the effects of nuclear weapons and in a broader context, the effects of nuclear war. Tsipis takes on an enormous task given the scope of the book, but manages to remain clear and (mostly) understandable throughout. He even does the math for you with numerous formulas (mostly in the appendices), but you can ignore all of them if math if not your strong suit, and still get the idea. Also covered are many of the so-called Star Wars weapons—"directed-energy" or laser-like devices. His chapter on the effects of nuclear war is sobering, but as grim as it is, it is probably too optimistic.

INDEX

M

N

O

P

R

S

BLUEPRINTS

With the exception of the missile schematic these are operational as built drawings. "Operational as built" means these blueprints reflect the details and dimensions used in actual construction, as opposed to those specified in the original contract drawings. Numerous differences between the two sets of drawings exist, generally due to last minute changes, and difficulties encountered by construction crews in the field.

CEBMCO stands for Corps of Engineers Ballistic Missile Construction Office.

Scales indicated on drawings do not apply to these reduced-sized reproductions.

The missile schematic drawing is probably an early research and development (R&D) drawing by the Glenn L. Martin Company. Several differences exist between the missile shown in this drawing and operational missiles. For example, the drawing shows three pitch rockets when only two were actually used. Additionally, the drawing shows the missile fitted with a destruct package. Destruct packages were installed only on R&D missiles and those used in operational test flights. There are also a few hand-written notes on the drawing.

FOR OFFICIAL USE ONLY

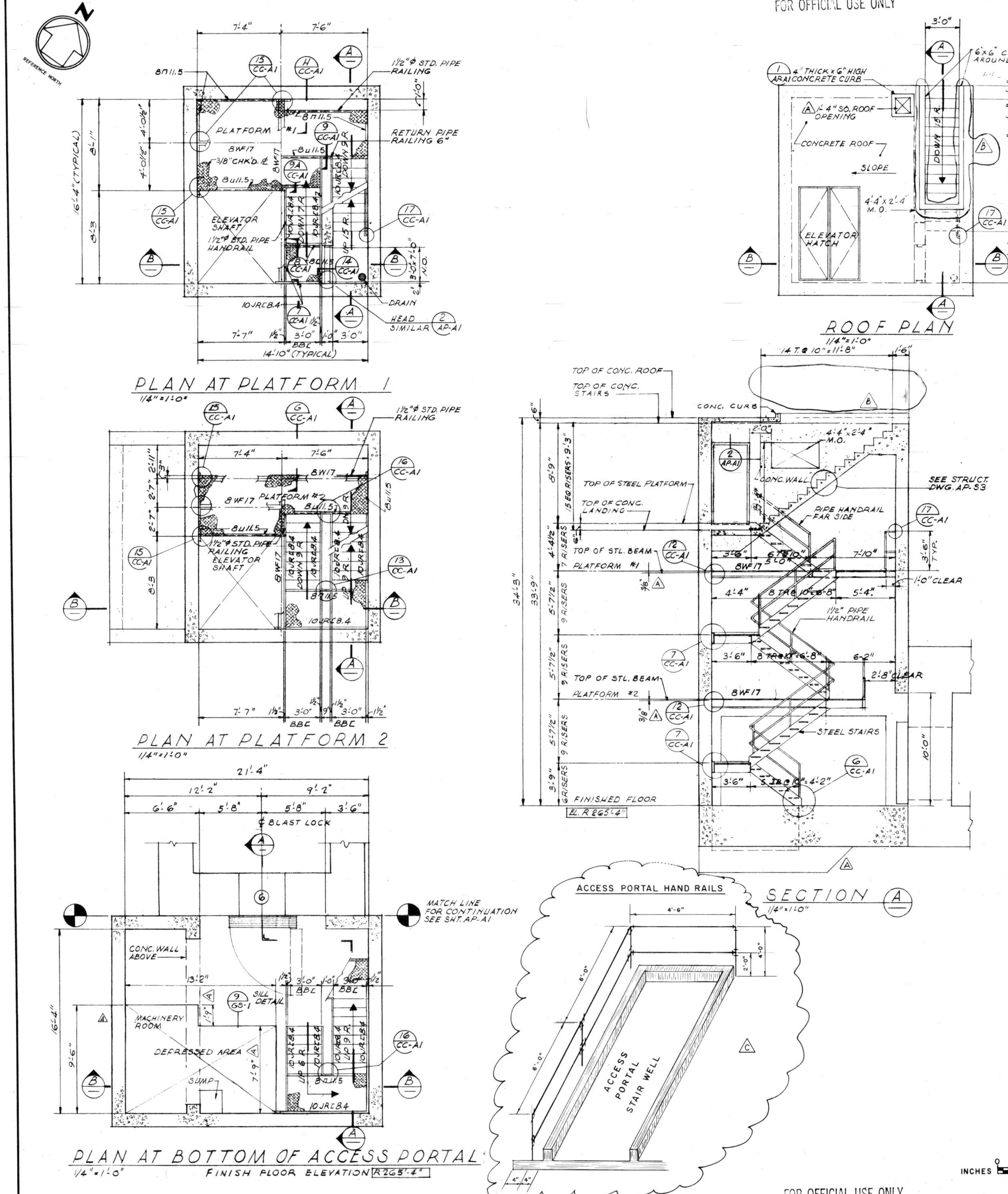

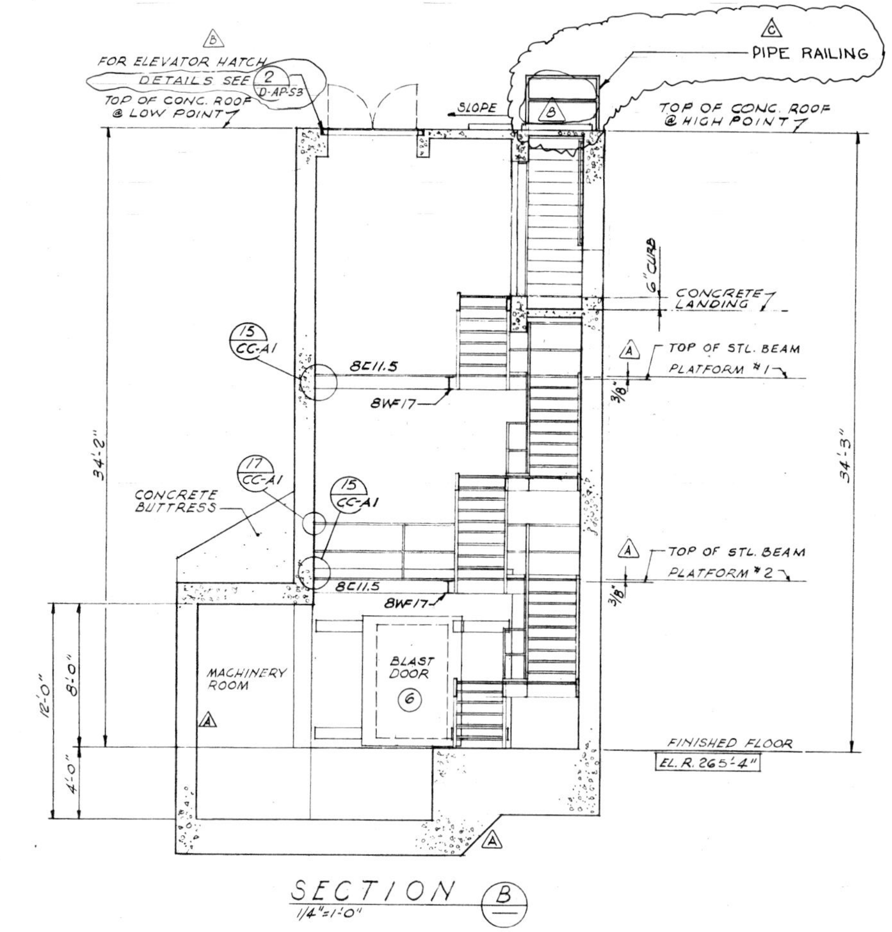

GENERAL NOTES

1. FOR DOOR SCHEDULE, SEE SHT. AP-A1
2. REFERENCE ELEVATION 300'-0" FOR ACTUAL ELEVATIONS, SEE CIVIL SHEETS.
3. FOR CONCRETE OPENINGS, WALL AND FOUNDATION DIMENSIONS, SEE STRUCTURAL SHEETS -SHT. AP-S3
4. ALL STEEL PLATFORM & LANDING DIMENSIONS SHOWN IN ELEVATION ARE TO TOP OF CHECKERED FLOOR PLATE.
5. MATERIAL FOR ACCESS PORTAL HAND RAILS-
   A -1-1/2" GALV. PIPE RAIL AND SCREWED FITTINGS.
   B - CIRCULAR FLOOR FLANGES WITH COUNTER SUNK HOLES AND ANCHOR BOLTS TO MATCH

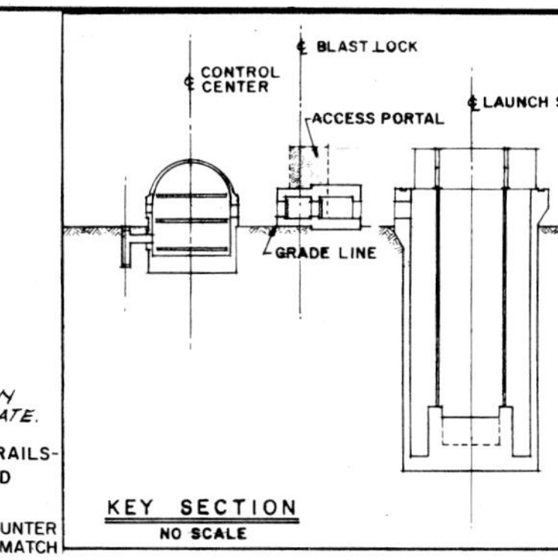

REVISION NO.	DATE	DESCRIPTION	AGENCY	BY
C	7 JAN 70	ADD HAND RAIL-PER DMAFB WO 1320-9	DMAFB	
O.A.B.	8-1-64	OPERATIONAL AS BUILT	AFBSD	SRC
C.A.B.	9-14-62	CONSTRUCTION AS BUILT SITE (S) # 1-18	AFBSD	am
—	6-15-62	ISSUED FOR CERTIFICATION		
B	8-25-61	REMOVED PIPE RAILING & REVISED ELEVATOR HATCH DETAILS.		
A	11-7-60	REVISIONS AS SHOW	AFBMD	

THE RALPH M. PARSONS COMPANY
ARCHITECT — ENGINEER
617 WEST SEVENTH STREET LOS ANGELES 17, CALIFORNIA

DEPARTMENT OF THE AIR FORCE
AIR FORCE BALLISTIC MISSILE DIVISION (ARDC)
INGLEWOOD, CALIFORNIA

DESIGNED BY E. NECHELES
DRAWN BY DICK, L.
CHECKED BY RB
SUBMITTED BY — DATE 9/23

A-E SHEET NO. S-AP-A2
A-E JOB NO. 1715-1C

SM-68B TECHNICAL FACILITIES
STANDARD PLANS

ACCESS PORTAL
PLANS & SECTIONS

APPROVED — DATE OCT 15 1960
MAJOR, USAF — OCT 15 1960
AFBMD — OCT 15 1960

AFBMD INDEX NO.

SPECIFICATION NO. ENG. 04-548-61 2
DATE OCT 15 1960 SCALE AS NOTED
DRAWING NUMBER CEBMCO SERIES 105/85
SHEET 85

FOR OFFICIAL USE ONLY

OPERATIONAL AS BUILT – NOT A CONTRACT DRAWING

FOR OFFICIAL USE ONLY

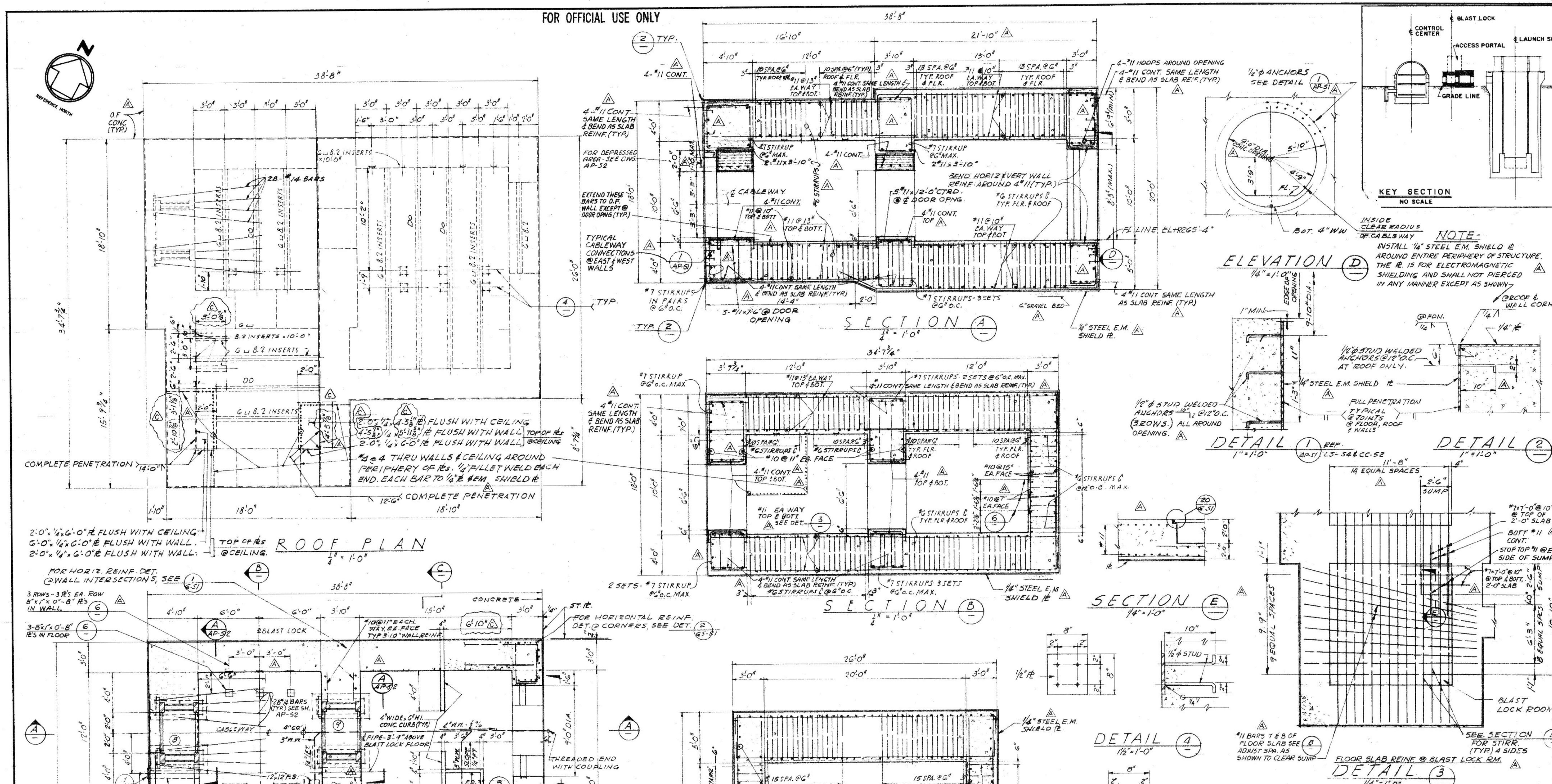

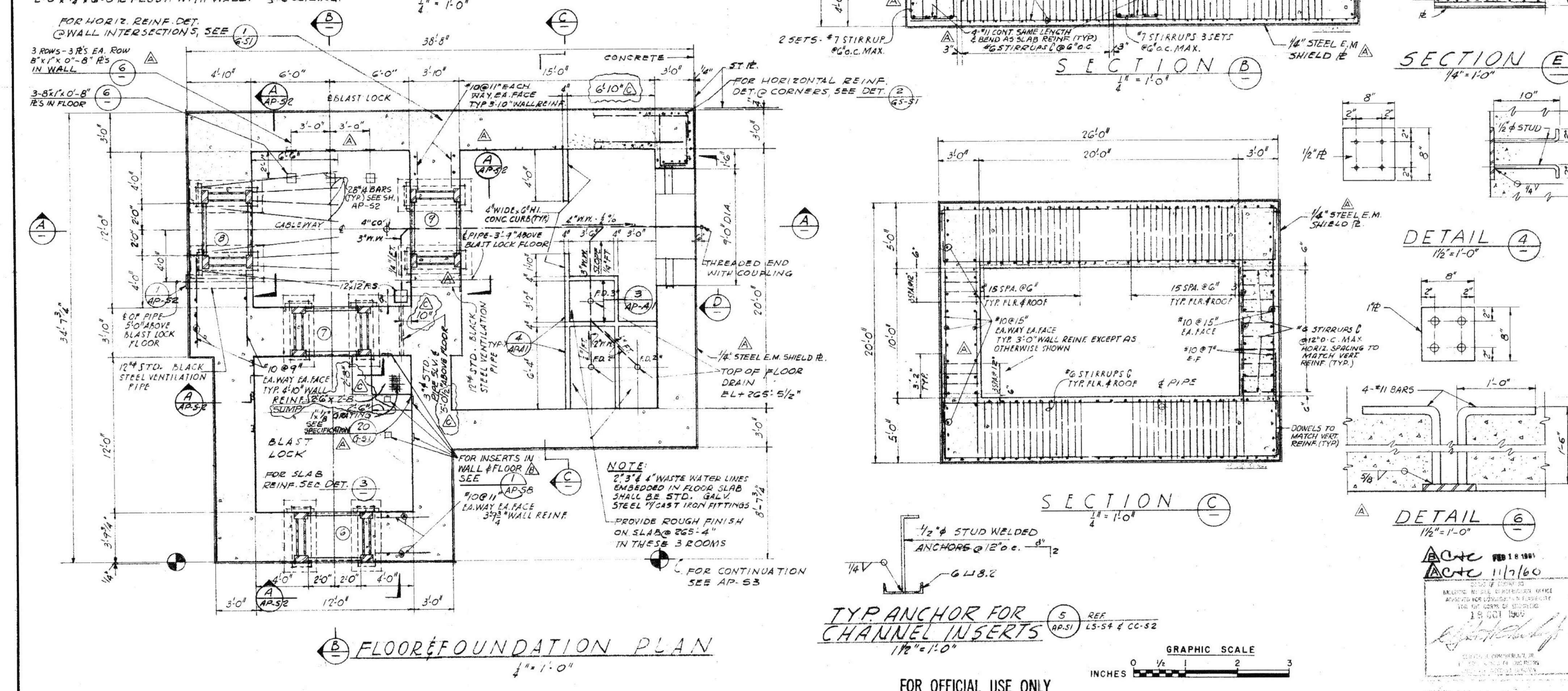

NOTES-

FOR ELECTRICAL CONDUIT SLEEVES THROUGH CONCRETE WALLS SEE DRAWING AP-E1 & DETAIL 7, SHEET G-S1

FOR ELECTRICAL GROUNDING, SEE DRAWING AP-E1.

FOR PIPE SLEEVES FOR PIPING, HEATING & VENTILATING, AND HYDRAULIC LINES SEE SHEET AP-S2

REVISION NO.	DATE	DESCRIPTION	AGENCY	BY
C.A.B.	9-14-62	CONSTRUCTION AS BUILT SITE (S) #9	AFBSD	QM
—	8-8-62	RE-ISSUED FOR CERTIFICATION SITE #9		
D	7-31-62	ADDED NEW AS BUILT DECAL		
—	6-15-62	ISSUED FOR CERTIFICATION - SITE 9		
C	6-15-62	REVISED PER "AS BUILT" MARKED PRINT		
B	2/14/61	INDICATED ℞ INSERTS IN FLOOR & WALLS. COC #577	AFBMD	TRF
A	11-7-60	REVISIONS AS SHOWN	AFBMD	TRF

THE RALPH M. PARSONS COMPANY
ARCHITECT — ENGINEER
617 WEST SEVENTH STREET LOS ANGELES 17, CALIFORNIA

DEPARTMENT OF THE AIR FORCE
AIR FORCE BALLISTIC MISSILE DIVISION (ARDC)
INGLEWOOD, CALIFORNIA

DESIGNED BY LEE
DRAWN BY HUGHES
CHECKED BY RWA
SUBMITTED BY [signature] DATE 9/23

A-E SHEET NO. S-AP-S1
A-E JOB NO. 1715-1C

SM-68B TECHNICAL FACILITIES STANDARD PLANS

ACCESS PORTAL AND BLAST LOCK STRUCTURES ROOF AND FLOOR PLAN AND SECTIONS

APPROVED: [signatures] OCT 1 5 1960; MAJOR, USAF OCT 1 5 1960; OCT 1 5 1960; AFBMD

SPECIFICATION NO. ENG. 04-548-61-2
DATE OCT 1 5 1960
SCALE AS NOTED
DRAWING NUMBER CEBMCO SERIES 105/86
SHEET 86

AFBMD INDEX NO.

FOR OFFICIAL USE ONLY

OPERATIONAL AS BUILT - NOT A CONTRACT DRAWING

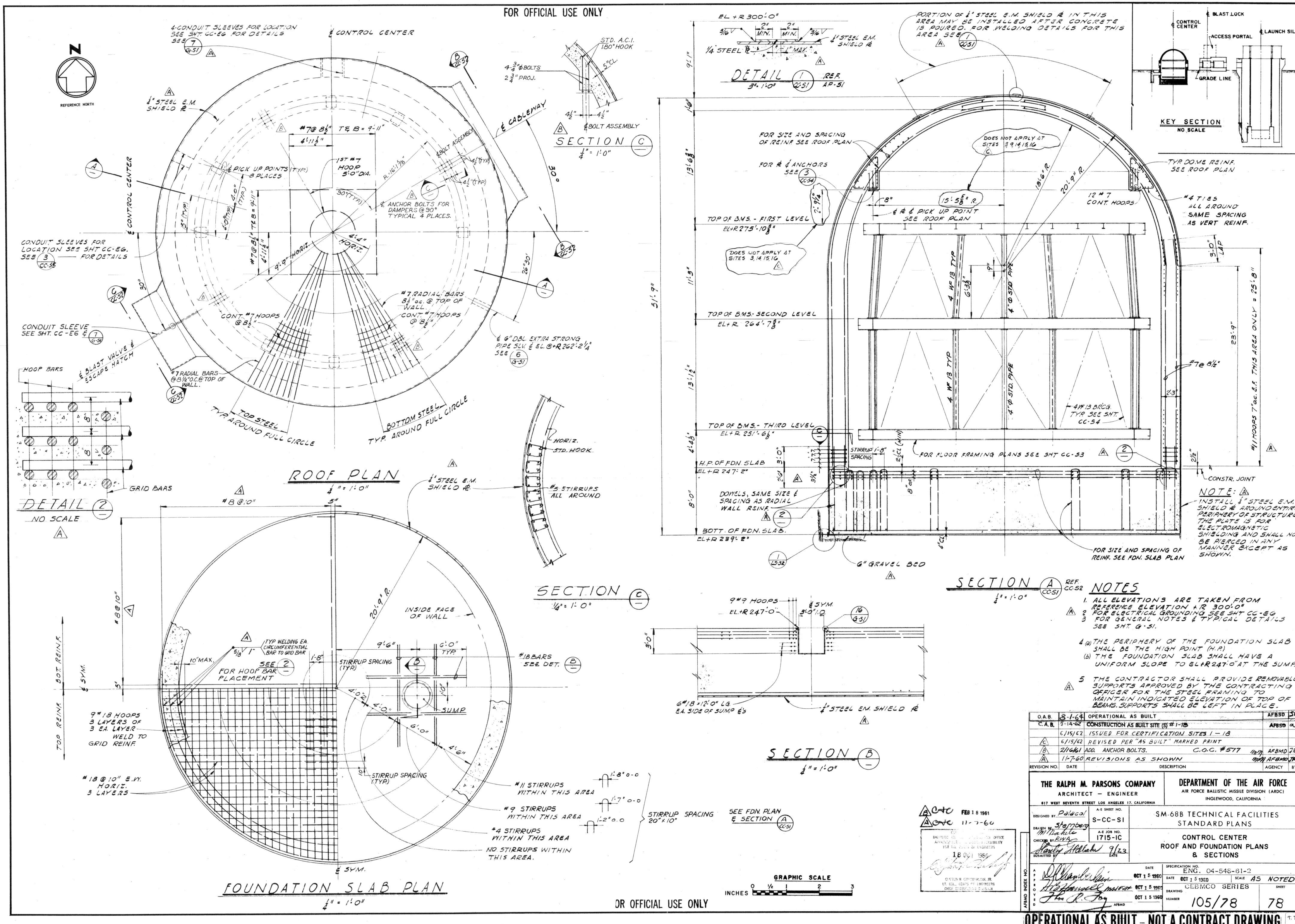

FOR OFFICIAL USE ONLY
REFERENCE NORTH
ROOF PLAN
1/4" = 1'-0"
℄ CONTROL CENTER
℄ CABLEWAY
TOP STEEL TYP. AROUND FULL CIRCLE
BOTTOM STEEL TYP. AROUND FULL CIRCLE
CONDUIT SLEEVES FOR LOCATION SEE SHT. CC-E6. SEE 3 CC-55 FOR DETAILS
CONDUIT SLEEVE SEE SHT. CC-E6
℄ BLAST VALVE & ESCAPE HATCH
#7 RADIAL BARS @ 8 1/2" O.C. @ TOP OF WALL.
CONT. #7 HOOPS @ 8 1/2
℄ 6" DBL EXTRA STRONG PIPE SLV. ℄ EL. @ +R 262'-2 1/4" SEE 6 G-51
1/4" STEEL E.M. SHIELD ℞
DETAIL 2
NO SCALE
HOOP BARS
GRID BARS
SECTION C
1/4" = 1'-0"
STD. A.C.I. 180° HOOK
BOLT ASSEMBLY
HORIZ.
STD. HOOK
#5 STIRRUPS ALL AROUND
FOUNDATION SLAB PLAN
1/4" = 1'-0"
#18 @ 10" E.W. HORIZ. 3 LAYERS
9 #18 HOOPS 3 LAYERS OF 3 EA. LAYER WELD TO GRID REINF.
FOR HOOP BAR PLACEMENT
INSIDE FACE OF WALL
STIRRUP SPACING (TYP)
SUMP
#18 BARS SEE DET. B
#11 STIRRUPS WITHIN THIS AREA
#9 STIRRUPS WITHIN THIS AREA
#4 STIRRUPS WITHIN THIS AREA
NO STIRRUPS WITHIN THIS AREA.
STIRRUP SPACING 20" × 10"
SEE FDN. PLAN & SECTION A CC-51
SECTION B
1/4" = 1'-0"
9 #9 HOOPS
EL +R 247'-0"
6 #18 × 12'-0" LG. EA. SIDE OF SUMP ℄'S
1/4" STEEL EM. SHIELD ℞
DETAIL 1 CC-51
3" = 1'-0"
REF. AP-51
PORTION OF 1/4" STEEL E.M. SHIELD ℞ IN THIS AREA MAY BE INSTALLED AFTER CONCRETE IS POURED. FOR WELDING DETAILS FOR THIS AREA SEE 1 CC-51
SECTION A CC-51
REF. CC-52
1/4" = 1'-0"
TOP OF BMS. - FIRST LEVEL EL +R 275'-10 3/8"
TOP OF BMS. - SECOND LEVEL EL +R 264'-7 3/8"
TOP OF BMS. - THIRD LEVEL EL +R 251'-6 3/8"
H.P. OF FDN SLAB EL +R 247'-2"
BOTT. OF FDN. SLAB. EL +R 239'-2"
6" GRAVEL BED
FOR FLOOR FRAMING PLANS SEE SHT CC-53
FOR SIZE AND SPACING OF REINF. SEE FDN. SLAB PLAN
TYP. DOME REINF. SEE ROOF PLAN
#4 TIES ALL AROUND SAME SPACING AS VERT REINF.
NOTE: INSTALL 1/4" STEEL E.M. SHIELD ℞ AROUND ENTIRE PERIPHERY OF STRUCTURE. THE PLATE IS FOR ELECTROMAGNETIC SHIELDING AND SHALL NOT BE PIERCED IN ANY MANNER EXCEPT AS SHOWN.
KEY SECTION
NO SCALE
℄ CONTROL CENTER
℄ BLAST LOCK
ACCESS PORTAL
℄ LAUNCH SILO
GRADE LINE
NOTES
1. ALL ELEVATIONS ARE TAKEN FROM REFERENCE ELEVATION +R 300'-0"
2. FOR ELECTRICAL GROUNDING SEE SHT. CC-E6
3. FOR GENERAL NOTES & TYPICAL DETAILS SEE SHT. G-51.
4 (a) THE PERIPHERY OF THE FOUNDATION SLAB SHALL BE THE HIGH POINT (H.P.)
(b) THE FOUNDATION SLAB SHALL HAVE A UNIFORM SLOPE TO EL +R 247'-0" AT THE SUMP.
5. THE CONTRACTOR SHALL PROVIDE REMOVABLE SUPPORTS APPROVED BY THE CONTRACTING OFFICER FOR THE STEEL FRAMING TO MAINTAIN INDICATED ELEVATION OF TOP OF BEAMS. SUPPORTS SHALL BE LEFT IN PLACE.
O.A.B. 8-1-64 OPERATIONAL AS BUILT AFBSD
C.A.B. 9-14-62 CONSTRUCTION AS BUILT SITE (S) # 1-18 AFBSD
6/15/62 ISSUED FOR CERTIFICATION SITES 1-18
6/15/62 REVISED PER "AS BUILT" MARKED PRINT
2/16/61 ADD. ANCHOR BOLTS. C.O.C. #577 AFBMD
11-7-60 REVISIONS AS SHOWN AFBMD
REVISION NO. DATE DESCRIPTION AGENCY BY
THE RALPH M. PARSONS COMPANY
ARCHITECT – ENGINEER
617 WEST SEVENTH STREET LOS ANGELES 17, CALIFORNIA
DEPARTMENT OF THE AIR FORCE
AIR FORCE BALLISTIC MISSILE DIVISION (ARDC)
INGLEWOOD, CALIFORNIA
A-E SHEET NO. S-CC-S1
A-E JOB NO. 1715-1C
SM-68B TECHNICAL FACILITIES
STANDARD PLANS
CONTROL CENTER
ROOF AND FOUNDATION PLANS & SECTIONS
SPECIFICATION NO. ENG. 04-546-61-2
DATE OCT 1 5 1960
SCALE AS NOTED
CEBMCO SERIES
DRAWING NUMBER 105/78
SHEET 78
GRAPHIC SCALE
INCHES 0 1/2 1 2 3
OR OFFICIAL USE ONLY
OPERATIONAL AS BUILT – NOT A CONTRACT DRAWING